Spectrum Requirement Planning in Wireless Communications

Wiley Series on Wireless Communications and Mobile Computing

Series Editors: Dr Xuemin (Sherman) Shen, *University of Waterloo, Canada*
Dr Yi Pan, *Georgia State University, USA*

The "Wiley Series on Wireless Communications and Mobile Computing" is a series of comprehensive, practical and timely books on wireless communication and network systems. The series focuses on topics ranging from wireless communication and coding theory to wireless applications and pervasive computing. The books offer engineers and other technical professionals, researchers, educators, and advanced students in these fields with invaluable insight into the latest developments and cutting-edge research.

Other titles in the series:

Misic and Misic: *Wireless Personal Area Networks: Performance, Interconnections and Security with IEEE 802.15.4*, January 2008, 978-0-470-51847-2

Perez-Fontan and Espiñeira: *Modeling the Wireless Propagation Channel: A Simulation Approach with Matlab*, August 2008, 978-0-470-72785-0

Ippolito: *Satellite Communications Systems Engineering: Atmospheric Effects on Satellite Link Design and System Performance*, September 2008, 978-0-470-72527-6

Myung: *Introduction to Single Carrier FDMA*, October 2008, 978-0-470-72449-1

Qian, Muller and Chen: *Security in Wireless Networks and Systems*, May 2009, 978-0-470-51212-8

Stojmenovic: *Wireless Sensor and Actuator Networks: Algorithms and Protocols for Scalable Coordination and Data Communication*, July 2009, 978-0-470-17082-3

Spectrum Requirement Planning in Wireless Communications

Model and Methodology for IMT-Advanced

Hideaki Takagi
University of Tsukuba, Japan

Bernhard H. Walke
RWTH Aachen University, Germany

John Wiley & Sons, Ltd

Other Wiley Editorial Offices

John Wiley & Sons Inc., 111 River Street, Hoboken, NJ 07030, USA

Jossey-Bass, 989 Market Street, San Francisco, CA 94103-1741, USA

Wiley-VCH Verlag GmbH, Boschstr. 12, D-69469 Weinheim, Germany

John Wiley & Sons Australia Ltd, 42 McDougall Street, Milton, Queensland 4064, Australia

John Wiley & Sons (Asia) Pte Ltd, 2 Clementi Loop #02-01, Jin Xing Distripark, Singapore 129809

John Wiley & Sons Canada Ltd, 6045 Freemont Blvd, Mississauga, ONT, L5R 4J3, Canada

Wiley also publishes its books in a variety of electronic formats. Some content that appears in print may not be
available in electronic books.

Library of Congress Cataloging-in-Publication Data

Spectrum requirement planning in wireless communications : model and methodology for IMT–Advanced /
edited by Hideaki Takagi, Bernhard H. Walke.
   p. cm.
Includes index.
ISBN 978-0-470-98647-9 (cloth)
1. Wireless communication systems–Standards. 2. Cellular telephone systems–Standards. 3. Mobile communication
systems–Standards. 4. Radio frequency allocation–International cooperation. I. Takagi, Hideaki.
II. Walke, Bernhard.
   TK5103.2.S74 2008
   621.384–dc22

                                                                                                2007049351

British Library Cataloguing in Publication Data
A catalogue record for this book is available from the British Library
ISBN 978-0-470-98647-9 (H/B)

Typeset by Sunrise Setting Ltd.
Printed and bound in Great Britain by Antony Rowe Ltd, Chippenham, England.

Contents

3   Spectrum Requirement Calculation for IMT-2000   45
Hideaki Takagi

4   Spectrum Requirement Calculation for IMT-Advanced   73
Marja Matinmikko, Jörg Huschke, Tim Irnich, Naoto Matoba, Jussi Ojala,
Pekka Ojanen, Hideaki Takagi, Bernhard H. Walke and Hitoshi Yoshino

About the Series Editors

Xuemin (Sherman) Shen (M'97-SM'02) received a B.Sc degree in electrical engineering from Dalian Maritime University, China in 1982, and M.Sc and Ph.D degrees (both in electrical engineering) from Rutgers University, New Jersey, USA, in 1987 and 1990, respectively. He is a Professor and University Research Chair, and the Associate Chair for Graduate Studies, Department of Electrical and Computer Engineering, University of Waterloo, Canada. His research focuses on mobility and resource management in interconnected wireless/wired networks, UWB wireless communications systems, wireless security, and ad hoc and sensor networks. He is a co-author of three books, and has published more than 300 papers and book chapters in wireless communications and networks, control and filtering. Dr Shen serves as a Founding Area Editor for IEEE Transactions on Wireless Communications; Editor-in-Chief for Peer-to-Peer Networking and Application; Associate Editor for IEEE Transactions on Vehicular Technology; KICS/IEEE Journal of Communications and Networks, Computer Networks; ACM/Wireless Networks; and Wireless Communications and Mobile Computing (Wiley), etc. He has also served as Guest Editor for IEEE JSAC, IEEE Wireless Communications, and IEEE Communications Magazine. Dr Shen received the Excellent Graduate Supervision Award in 2006, and the Outstanding Performance Award in 2004 from the University of Waterloo, the Premier's Research Excellence Award (PREA) in 2003 from the Province of Ontario, Canada, and the Distinguished Performance Award in 2002 from the Faculty of Engineering, University of Waterloo. Dr Shen is a registered Professional Engineer of Ontario, Canada.

Dr Yi Pan is the Chair and a Professor in the Department of Computer Science at Georgia State University, USA. Dr Pan received his B.Eng and M.Eng degrees in computer engineering from Tsinghua University, China, in 1982 and 1984, respectively, and his Ph.D degree in computer science from the University of Pittsburgh, USA, in 1991. Dr Pan's research interests include parallel and distributed computing, optical networks, wireless networks, and bioinformatics. Dr Pan has published more than 100 journal papers with over 30 papers published in various IEEE journals. In addition, he has published over 130 papers in refereed conferences (including IPDPS, ICPP, ICDCS, INFOCOM, and GLOBECOM). He has also co-edited over 30 books. Dr Pan has served as an editor-in-chief or an editorial board member for 15 journals including five IEEE Transactions and has organized many international conferences and workshops. Dr Pan has delivered over 10 keynote speeches at many international conferences. Dr Pan is an IEEE Distinguished Speaker (2000–2002), a Yamacraw Distinguished Speaker (2002), and a Shell Oil Colloquium Speaker (2002). He is listed in Men of Achievement, Who's Who in America, Who's Who in American Education, Who's Who in Computational Science and Engineering, and Who's Who of Asian Americans.

Preface

People drive cars without knowing the mechanics and electronics inside. Likewise people will exchange video mails over mobile phones without bothering about digital coding, channel assignment, routing, and so on by taking advantage of the ever progressing wireless communication technologies. However, there is a certain group of people behind the scenes who are concerned and working hard to maintain and develop the systems for the sake of all other people. Are you interested (technically) in knowing what roads and highways for voice or data messages from your cell phone use to reach your friends, though you do not see any apparent passageway out of your phone? Then this book is for you!

Due to the continuous development in semiconductor technology, processing units and memory are no longer bottlenecks in computers. Similarly, the explosive growth in the capacity of optical fiber, the bandwidth of wireline communication is virtually unlimited in communication networks. Unfortunately, this is not the case in wireless communication networks. The propagation medium of electromagnetic waves used for wireless communication is the *atmosphere* endowed by Nature. The spectrum bandwidth that can be used for public terrestrial and satellite communications is restricted for physical reasons (attenuation of propagating radio waves, absorption of radio signal energy by gases, water vapor and rain, absorption by multipath propagation and reflection of waves at obstacles) as well as for manmade reasons (e.g. reservation of spectrum for radio broadcasting or military use). Only a few communication channels are available at a given time and place from a given frequency band. While considerable efforts are being dedicated to exploring higher and higher frequency spectrum for mobile use, we have to divide the available spectrum bandwidth, which is a limited resource, among the services (e.g. mobile radio, broadcasting, public authority, military, radar navigation and surveillance). Today the worldwide usage of radio frequencies is administered by the *International Telecommunication Union* (ITU), a United Nations organization. The set of *Radio Regulations* of the Radiocommunication sector of ITU (ITU-R) is a binding intergovernmental treaty governing the use of radio spectrum.

We are now (the year 2008) in the Third Generation (3G) of mobile communication standards and technology. The 3G mobile services started around the early 2000s to serve voice and Internet access. Initially the mean user date rate was 384 kbit/s, and it has been substantially increased to a few Mbit/s. Two notable examples of 3G radio interface are wideband code division multiple access (W-CDMA) or Universal Mobile Telecommunications System (UMTS) developed by the Third Generation Partnership Project (3GPP) and cdma2000 developed by 3GPP2. International Mobile Telecommunications-2000 (IMT-2000) is the 3G mobile system specified by ITU-R. The 3GPP/3GPP2 have developed their evolutional systems, which are referred to as *future development of IMT-2000* in ITU-R documents. The Fourth Generation (4G) systems, called 'systems beyond IMT-2000' in some early ITU-R documents, are now called *IMT-Advanced*. We will use the ITU terminology IMT-2000, future development of IMT-2000 and IMT-Advanced in this book.

IMT-Advanced systems are currently being specified by the ITU-R. Until October 2007, IMT issues were under the responsibility of the Working Party 8F (WP8F) of ITU-R. This book addresses part of the work of the WP8F. IMT-Advanced systems are expected to be in operation around the year 2015. They will support a rich variety of mobile services and applications with a peak user data rate of 100 Mbit/s for highly mobile terminals and up to 1 Gbit/s for slowly moving terminals in metropolitan areas. These extremely high data rates will require much additional frequency spectrum, compared to the current allocations for IMT-2000 mobile services. Therefore, the question 'how much spectrum will be needed in IMT-Advanced systems' is the main subject of this book.

The ITU has developed approaches for estimating the spectrum requirements of wireless systems. The ITU methodology for estimation of spectrum requirements for IMT-2000 systems is given in Recommendation ITU-R M.1390. Report ITU-R M.2023 provides the estimate of spectrum requirements for terrestrial components of IMT-2000. The results were used as input to the World Radiocommunication Conference 2000 (WRC-2000), which identified additional spectrum bands for IMT-2000 systems to complement the spectrum identified initially for IMT-2000 by the World Administrative Radiocommunication Conference 1992 (WARC-92). The methodology in Rec. ITU-R M.1390 considers a single network with service delivery based on a circuit-switched radio network. However, according to the framework and objectives for IMT-Advanced shown in Recommendation ITU-R M.1645, the service delivery in the future is based mainly on broadband packet-switched radio networks supporting the Internet Protocol (IP). The seamless interworking between different radio access systems, namely Pre-IMT, IMT-2000, future development of IMT-2000 and IMT-Advanced systems, as well as such wireless systems as specified by IEEE Project 802, is required as they are expected competitively to offer their services to users in the same location. It appears to be extremely difficult to predict the frequency spectrum needs of such a scenario what with all these mobile and wireless systems requiring a share of the spectrum in order to be able to generate business.

Therefore, ITU-R has developed a new methodology to calculate the radio spectrum requirements for IMT-Advanced systems based on market survey data predicting the traffic load for the year 2010 and beyond. The new methodology was approved as Recommendation ITU-R M.1768 in June 2006. By using the new methodology, ITU-R has calculated the spectrum requirements for IMT-Advanced as shown in Report ITU-R M.2078. These were provided to the World Radiocommunication Conference 2007 (WRC-07) held in October–November 2007 in Geneva, Switzerland. There, a set of new spectrum allocations for Mobile Service and identifications for IMT-Advanced was determined by taking the needs of various regions in the world into account (some of which have a predominant need for satellite-based communication). The calculation algorithm embedded in the methodology is complex with detailed equations, and it is implemented only in a tool package whose structure and contents appear rather complicated. Thus it is difficult to know the internal model and methodology used in the calculation. It will be very hard for outsiders to understand the reasons for the decisions made at WRC-07.

This volume presents a *self-contained handbook* of the model and methodology used for the spectrum requirement calculation for the IMT-Advanced systems and also reports on the methodology used for the previous IMT-2000 system. It shows the underlying theoretical models as well as the derivation of the mathematical formulas, which do not appear explicitly in the above-mentioned ITU-R documents. Therefore, the reader can learn how the spectrum requirement is calculated for real systems, which will prevail worldwide in the forthcoming era of ubiquitous computing and communication. In addition, the authors hope that the book provides the *base camp* possibly to develop a further advanced methodology to be applied for WRC decisions in the years 2011 or later. The contributors to the book are the members of the mobile IT Forum (mITF) in Japan and the WINNER project partners in Europe (see below) who, themselves, actually developed the new methodology for WP8F of ITU-R.

Although there are ten authors, this handbook is not just a disorderly collection of independently contributed chapters. Instead, the editors have tried to correlate the chapters consistently and unify the notation of symbols throughout. Readers are expected to read the book through to understand the main characteristics of the worldwide leading mobile/wireless communication systems and the methodology for spectrum requirement calculation systematically, mathematically and historically.

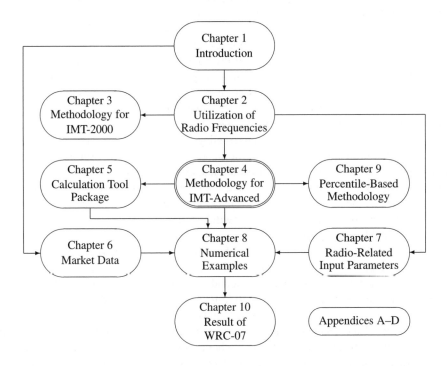

This book consists of ten chapters and four appendices. The relationship between the chapters is shown in a flow chart style above. The main stream starts with Chapter 1 and goes through Chapters 2, 4, 8 and 10. Among them Chapter 4 is the core of the book. Chapters 3, 5, 6, 7 and 9 and Appendices A–D provide supplementary but useful information.

We begin in Chapter 1 with an introduction to the recent trends in mobile communication systems with respect to applications and services as well as radio interface technologies. We also touch upon relevant standardization activities by various organizations. Chapter 2 presents a basic knowledge of the utilization of the spectrum of electromagnetic waves for communication systems. This chapter helps the reader to understand how the spectrum and radio services/systems are managed by ITU-R. We mention both terrestrial services (fixed, mobile and broadcasting) and satellite services (fixed and mobile). We then explain technologies for radio communication systems such as wireless local area networks, terrestrial digital broadcasting and short-range communications, e.g. Bluetooth. In particular, cellular systems are described in detail as they offer a seamless service to mobile users, and their spectrum requirement is the principal topic of this book. These two chapters should be useful as the background to the spectrum requirement planning in wireless communications.

In Chapter 3, we review the spectrum requirement calculation methodology for 3G mobile systems denoted as IMT-2000, focusing on the years from 2000 to 2010, as given in Recommendation ITU-R M.1390. Numerical examples of calculation are shown in parallel from Report ITU-R M.2023. At the time of this estimation (late 1990s), main applications of mobile communications were voice, facsimile and some multimedia applications. The methodology was based on the conventional Erlang-B and Erlang-C approaches for circuit-switched traffic. The methodology and numerical results were presented at the WRC-2000, where the additional spectrum bands were identified for IMT-2000.

The spectrum requirement calculation for IMT-Advanced is given in Chapter 4, which explains a new spectrum estimation methodology developed by ITU-R, focusing on the years 2010 and onwards. This chapter, the core of the book, is based on Recommendation ITU-R M.1768. After an overview of the development story, we introduce models and input parameters used in the methodology such as service categories, service environments and radio environments. Service environments are combinations of teledensities (dense urban, suburban and rural) and service usage patterns (home, office and public area). Radio environments refer to types of cell such as macro cell, micro cell, pico cell and hot spot. Our algorithm starts with the calculation of traffic demands from market data for different service categories in different service environments and its distribution to radio environments for each radio access technique group. We then proceed to calculate the system capacity (bit/s) so as to satisfy the requirement of the quality of service (QoS), which is given in terms of the maximum allowable blocking probability for circuit-switched service categories or in terms of the maximum allowable mean packet delay for packet-switched service categories. To do so, we use the multidimensional Erlang loss model for circuit-switched service categories and the M/G/1 nonpreemptive priority queueing model for packet-switched service categories. These models enable category-dependent treatment and statistical multiplexing of different service categories. The capacity requirement per cell (bit/s/cell) is simply divided by the area spectral efficiency factor to give the spectrum requirement (Hz). The resulting spectrum requirements are then aggregated and necessary adjustment is made to yield the total spectrum requirement.

The spectrum requirement calculation methodology for IMT-Advanced has been implemented into a tool package on MS Excel worksheets, which is described in detail in Chapter 5. The tool is now publicly available on the ITU web site 'http://www.itu.int/ITU-R/study-groups/docs/speculator.doc'. Chapter 6 reports the analysis of market data collected from responses to the questionnaire on services and market for IMT-Advanced and other radio systems distributed worldwide on the basis of Report ITU-R M.2072. It also details the calculation of market study parameter values for the input to the spectrum calculation methodology. Chapter 7 explains radio access technique group (RATG) approach and radio-related parameters used for the traffic demand and distribution calculation (cell area, application data rate, etc.) and those for the spectrum requirement calculation (spectral efficiency, number of overlapping networks, minimum deployment per operator, etc.) in the methodology. This chapter is based on Report ITU-R M.2074.

Chapter 8 presents numerical examples of the spectrum estimation, based on Report ITU-R M.2078, by using the calculation tool described in Chapter 5, the example input market parameter values shown in Chapter 6 and the example input radio parameter values shown in Chapter 7. This illustration should help readers to gain a clear view of the methodology. Chapter 9 gives a possible extension to the current methodology to meet the user requirement with respect to the delay percentile for packet-switched service categories (the user requirement assumed in the methodology of Chapter 4 was the mean packet delay). This extension is not included in the ITU-R documents, but it may suggest future research direction. Finally, Chapter 10 tells you what the methodology has finally brought us. The result of WRC-07 in terms of spectrum identification for IMT-Advanced is reported.

We provide several appendices for the convenience of the reader. In Appendix A, we derive queueing theory formulas used in the methodologies for IMT-2000 (Chapter 3) and IMT-Advanced (Chapter 4). Appendix B shows an illustrative set of market study parameter values which was actually used by ITU-R for the preparation of WRC-07 (errors in numerical values in Report ITU-R M.2078 have been corrected). Appendix C provides a list of acronyms and symbols used in the book. Appendix D lists relevant ITU-R documents and web sites. The bibliography is limited to those books, technical papers and ITU-R documents that are directly related to the topic of this book. The subject index refers only to key words for the methodology.

The target readership of this book is engineers of mobile communication operators and vendors and of national and regional radio regulators of the Administrations who are not familiar with teletraffic and queueing theory and its applications. These engineers should know and are expected to appreciate the simple underlying models and mathematical formulas used in the software package that produces numerical values for their system design. Other groups may be researchers of queueing theory who are not familiar with practical applications, and graduate students in the fields of applied probability, operations research, telecommunications and networking engineering. Queueing theory specialists would be delighted to find that the basic classic formulas are useful in modern communication systems.

This book is realized through the joint efforts of the following ten co-authors:

Mitsuhiro Azuma, Fujitsu Laboratories, Japan

Jörg Huschke, Ericsson, Germany

Tim Irnich, RWTH Aachen University, Germany

Marja Matinmikko, VTT Technical Research Centre of Finland, Finland

Naoto Matoba, NTT DoCoMo, Japan

Jussi Ojala, Nokia Corporation, Finland

Pekka Ojanen, Nokia Corporation, Finland

Hideaki Takagi, University of Tsukuba, Japan

Bernhard H. Walke, RWTH Aachen University, Germany

Hitoshi Yoshino, NTT DoCoMo, Japan

In each chapter, the names of contributors to the chapter are listed. The chapter editor comes first, followed by other contributors in alphabetical order of their last names (our apology to Dr Yoshino who always comes last). These authors not only contributed the manuscripts but also helped the editors to put them together into a complete book. We are also happy to announce that doctoral degrees were awarded to Mitsuhiro Azuma (2007) and Tim Irnich (2008) and a licentiate degree to Marja Matinmikko (2007) while they were working on the spectrum requirements of IMT-Advanced.

The work of the European contributors was mostly carried out in the Wireless World Initiative New Radio (WINNER) and the WINNER II projects, which were partially funded by the European Union. The WINNER project partners participated in the ITU-R preparations towards the WRC-07 on the spectrum requirements of IMT-Advanced by preparing a number of contributions via the European Conference of Postal and Telecommunications Administrations (CEPT) with a view to forming common European contributions to the ITU-R. A number of contributions were also submitted directly to the ITU-R. Many of the WINNER contributions were targeted to the development of the spectrum calculation methodology as well as defining the different input parameters to the methodology. In essence, most parts of the spectrum calculation methodology originate from the WINNER project, including the software tool that was used to derive the final spectrum requirements. The work of the WINNER project partners was done in close cooperation with the mobile IT Forum (mITF) from Japan.

Funding towards developing the methodology by the Federal Minister of Research and Education in Germany in project 4GSpectrum is worth noting. The long-term research funding of VTT Technical Research Centre of Finland on 4G systems is also acknowledged.

Mobile IT Forum (mITF) is a Japanese forum which has been performing research on future mobile communication systems and services. In response to the start of ITU-R study on the methodology for calculation of spectrum requirements for IMT-Advanced, the mITF created an ad hoc group with an aim to providing technical basis for Japanese national preparation. The ad hoc group prepared many technical documents which were finally approved as Japanese contributions to the ITU-R. The members of the ad hoc group actively participated in the ITU-R activity for the development of the spectrum calculation methodology and enjoyed close interaction with WINNER colleagues. The mITF contributed to the development of a new methodology using a multidimensional Erlang-B formula for circuit-switched service categories in addition to the refinement of the methodology using an M/G/1 nonpreemptive priority queueing model for packet-switched service categories which was originally proposed by the WINNER project members.

Hideaki Takagi's work was supported by the mITF through Commissioned Research ACA16104 and ACA17044 in 2004–2005 as well as by the Grant-in-Aid for Scientific Research from the Ministry of Education, Science, Sports and Culture of Japan in 2006–2007.

The continuous support for the development of the spectrum estimation methodology by Dr Werner Mohr, Project Coordinator of the WINNER projects, is also worth mentioning and is appreciated by the editors.

Hideaki Takagi,
University of Tsukuba, Japan

Bernhard H. Walke,
RWTH Aachen University, Aachen, Germany

1

Introduction

Bernhard H. Walke and Hitoshi Yoshino

This chapter reviews trends in mobile communications and spectrum usage and explains why and how we need spectrum allocation to IMT systems.

1.1 Trends in Mobile Communication

This section describes trends in mobile communications on three fronts; applications and services, radio interface technologies and standardization.

1.1.1 Mobile applications and services

User expectations are increasing to support a wide variety of applications and services in mobile communications after the advent of broadband Internet access in wired communication. In the near future, wireless and mobile technology will play a vital role in providing 'continuous connectivity' between (end user) terminals and a variety of services. Note that *mobile* systems support an application running on the user terminal without interruption, even when moving with high mobility. *Wireless* systems connect slow-moving terminals to the Internet, interrupting service when switching between network access points. In a scenario where 'everybody and everything is always connected to access personalized services', several types of 'human to human', 'human to machine' and 'machine to machine' communication link can exist (Walke and Kumar 2003). See Figure 1.1.

The majority of presently used 'human to human' information exchange is voice based. A clear shift towards data services is observed. In 'human to machine' and 'machine to machine' interaction, the volume of information exchanged is small and a short duration 'session' at a low data rate is sufficient in most cases. For 'human to human' and 'machine to human' interaction for work or leisure, the opposite applies with long session duration and a high data rate required.

Spectrum Requirement Planning in Wireless Communications   Edited by Hideaki Takagi and Bernhard H. Walke

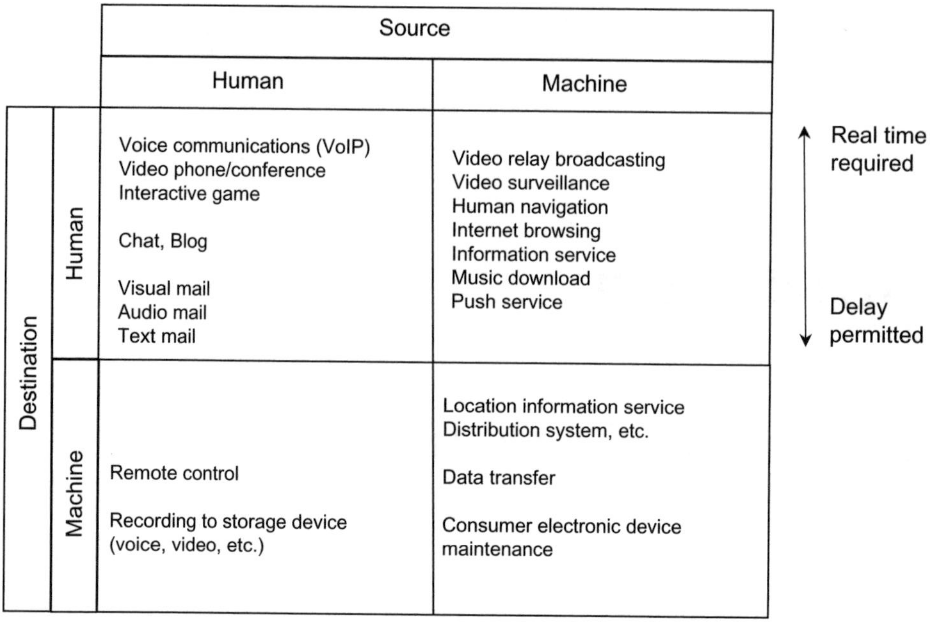

Source		
Human	**Machine**	
Voice communications (VoIP) Video phone/conference Interactive game Chat, Blog Visual mail Audio mail Text mail	Video relay broadcasting Video surveillance Human navigation Internet browsing Information service Music download Push service	Real time required Delay permitted
Remote control Recording to storage device (voice, video, etc.)	Location information service Distribution system, etc. Data transfer Consumer electronic device maintenance	

(Row labels: Destination — Human (top row), Machine (bottom row))

Figure 1.1 Wireless communication applications to support real-time and nonreal-time future wireless services.

Intelligent spaces in the future wireless world shall contain myriads of 'intelligent' wireless devices such as sensors and actuators embedded in appliances and/or carried by humans and interacting with each other as well as with their physical environment. There, the spontaneous information exchange may be based on dynamically configurable ad hoc networks of very low-power transceivers located in devices with varying information-processing capabilities. The transceivers might be connected to sensors and/or actuators, such as microphones and speakers. The very high concentration of such transceivers and the need to communicate not only in short range but also over medium to large distances would need a large spectrum bandwidth. Also, some of the envisaged future wireless and mobile applications and services will be 'location aware'. This requires suitable new air interface technology capable of combining the functions of data transmission with those of precise localization and position tracking.

The traffic resulting from data communication-based applications and services is similar to that known from the Internet. Accordingly, a packet-based delivery over radio is appropriate. The traffic flow may be unidirectional from transmitter to receiver terminal or bi-directional. The flow may be either symmetrical or asymmetrical, and the service required by an application may be real-time or nonreal-time oriented. The digitized and packetized information transmission permits an integration and convergence of technologies known from information science, telecommunications and contents provisioning. The wireless traffic amount resulting from all the three domains is increasing and consequently consumes ever increasing spectrum bandwidth.

Moore's law[1] appears also to apply for the bandwidth consumption. The end users tend to embrace applications and services utilizing an ever faster data rate. The service data rate (offered/required) is doubling every 12 months or so. The constant increase in 'users' injects further positive feedback into the system, thus sending the frequency bandwidth demand to an exponential increase.

This potential growth scenario requiring more and more bandwidth to allow a steady further development of wireless and mobile systems can be sketched in the light of the following:

- The extent of good quality radio coverage is inversely proportional to the transmitted data rate. The cost of 'continuous' and 'all time everywhere' radio coverage increases very sharply with the transmitted data rate.

- The higher the service data rate, the larger is the required bandwidth and the higher is the frequency range where some additional spectrum might be available.

- Deregulation policies of regulators, aiming at competition between operators, result in fragmentation of frequency spectrum licensed to mobile operators, inversely affecting spectrum-efficient use of a radio band allocated for mobile services.

- The present users (systems and service providers) of the already allocated frequency bands would like to make the most out of their allocation. The introduction of sophisticated mechanisms in the standardized air interfaces, e.g. space-time coding, smart antenna systems and multihop links to improve the radio coverage, appear to be a direct consequence of frequency spectrum shortage for mobile radio use.

- Although the spectrum efficiency of radio systems is continuously increasing, much more spectrum is required, in general. Moreover, additional frequency spectrum will be necessary in the low frequency range in order to provide the required coverage in wide areas.

- The variety of networks for provision of seamless services in private to public and short range localized coverage to wide area coverage will find its limits by the demand for cost effectiveness of the corresponding business cases.

The globalization of markets requires a very wide consensus going beyond technology standardization. Especially, the interworking of permanently established and spontaneously created networks shall be fostered to improve user acceptance on mobile/wireless services.

1.1.2 Radio interface technologies

History of mobile radio systems before IMT-2000

The first generation (1G) of wireless technology, dedicated to telephony, started in the 1980s with the analog cellular phone standards. There, mobile terminals and base stations use analog signal processing to transmit and receive the radio signals that propagate in any

---

[1]Moore's law generally refers to a trend that the capability of electronic devices grows at an exponential rate. The observation was first made by Intel co-founder Gordon E. Moore in his paper published in 1965 with respect to the number of transistors on an integrated circuit chip.

generation of mobile systems as analog signals through the atmosphere. Examples are the Advanced Mobile Phone Service (AMPS) deployed in the United States, the Nordic Mobile Telephone (NMT) in Scandinavian countries, the Netherlands and Switzerland, RC2000 in France, Total Access Communication System (TACS) in the United Kingdom, and C450 in Germany and Portugal. These continued until being replaced in the mid-1990s by the second generation (2G) technology that is based on digital signal processing applied in base stations and mobile terminals. The 2G services, called Personal Communications Service (PCS) in the United States, comprise mobile voice and narrowband data communication. The systems use combinations of multiplexing techniques at the air (radio) interface such as frequency division multiplex (FDM), time division multiplex (TDM) and code division multiplex (CDM) combined with the respective access protocols, namely frequency division multiple access (FDMA), time division multiple access (TDMA) and code division multiple access (CDMA). The 2G systems worth mentioning are Global System for Mobile communications (GSM) standardized by the European Telecommunication Standards Institute (ETSI) that reached a 75% market share worldwide, IS (Interim Standard)-95/cdmaOne according to the Telecommunications Industry Association (TIA) in the United States, and Personal Digital Cellular (PDC) specified by the Research Center for Radio (RCR) in Japan. An evolutionary technology called 2.5G introduced multiplexing of data packets to a common radio channel for mobile Internet access at 128 kbit/s mean transmission rate. Worth mentioning are GSM/EGPRS (Enhanced General Packet Radio Service) and the evolution technology of the cdmaOne system that increased the peak user data rate to 256 kbit/s.

Capability of future mobile and wireless systems

The framework and overall objectives of the future development of IMT-2000 and IMT-Advanced are described in Recommendation ITU-R M.1645, which was approved by ITU-R in June 2003. Figure 1.2 shows the capabilities of mobile and wireless systems, which are envisaged in Recommendation ITU-R M.1645.

Due to the wide spread of mobile Internet access supporting a wide variety of data rates and a wide range of mobility, current mobile systems such as IMT-2000 have evolved by the addition of more and more capabilities. Future broadband mobile Internet access will require a new mobile access and new nomadic/local area wireless access technologies. It is envisaged that those new technologies will need to support data rates of up to approximately 100 Mbit/s for high mobility and up to approximately 1 Gbit/s for low mobility, judging from broadband applications currently available in wired networks. To meet the high aggregate data rate requirements, IMT-Advanced systems will require considerably wider bandwidths than current mobile communications systems. Even if the spectral efficiency of the IMT-Advanced system will be considerably higher than in current systems, IMT-Advanced systems will require bandwidths of up to 100 MHz to support aggregate data rates of up to 1 Gbit/s. Currently existing bands for IMT-2000 are too narrow and fragmented, and this does not allow the implementation of 100 MHz carriers. Therefore, the deployment of IMT-Advanced systems with its fully envisioned capabilities is not possible on existing bands.

A similarity of applications and services across different wireless systems stimulates the convergence and interwork of the wireless systems. The prevalence of IP-based applications accelerates this convergence and interwork of the telecommunication systems.

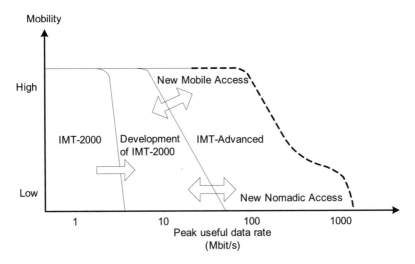

Figure 1.2 Capabilities of mobile systems ('Van Diagram').

Experience from the past has shown that the idea of a universal system able to cover all the needs of wireless and mobile applications cannot be realized. Instead, a multitude of air interfaces has been standardized and will continue to grow in the future to cover the specifically different needs of mobile and wireless communicating users in the various usage scenarios.

Mobile systems such as GSM, shown in Figure 1.3, together with its General Packet Radio Service (GPRS) and its evolution called Enhanced Data Rate for Global Evolution (EDGE), are covering the full range of mobility from fixed to high speed train mobility. The GSM/GPRS/EDGE is supporting low mobile data rates only and is, currently, dominating the world in 2G systems. The CDMA 1x, another 2G system with similar throughput capacity and mobility support, is also shown in the figure. Third generation (3G) systems such as Universal Mobile Telecommunications System (UMTS) and its evolutions called High Speed Downlink Packet Access (HSDPA) and High Speed Uplink Packet Access (HSUPA), in short, High Speed Packet Access (HSPA), and a technology called CDMA EV-DO (Evolution Data Only), offer a substantially increased throughput for the full range of terminal mobility. These systems are being planned to evolve further towards mobile broadband supporting systems, shown in Figure 1.3 as UMTS-LTE (Long Term Evolution) and Ultra Mobile Broadband (UMB). The respective standardization processes have started already. As can be seen in Figures 1.2 and 1.3, the right-hand upper corner is difficult to cover, limiting mobile broadband use to moderate speed of terminal movement.

Cordless technology such as Digital Enhanced Cordless Telecommunications (DECT) and Personal Handy-phone System (PHS) covers wireless telephony and low rate data, supporting only slow-moving terminals that are close (approximately 50 m) to the serving base station. Bluetooth is specialized to cover voice and data in the personal area of a human, bridging typically up to 10 m only. Wireless systems have been standardized by Project 802 of the Institute of Electrical and Electronics Engineers (IEEE). According to the standard IEEE 802.11, the wireless local area network (WLAN) is intended to serve nomadic terminals

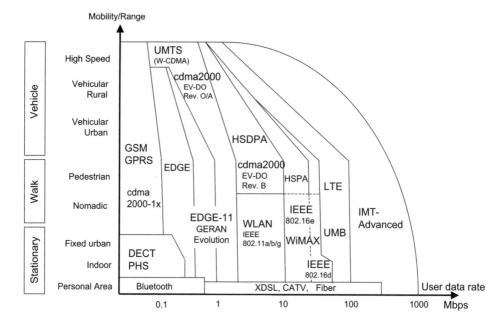

Figure 1.3 Terminal mobility versus peak data rate supported by wireless/mobile systems. (Reproduced by permission of © 2006 John Wiley & Sons, Ltd.)

with connectivity over radio to the Internet. The metropolitan area network (MAN) standard IEEE 802.16, called the Worldwide Interoperability for Microwave Access (WiMAX), was originally aimed to connect fixed subscriber stations via radio to a base station, but evolved to support terminals moving at vehicular speed. The system is expected to evolve further towards a mobile broadband system supporting the full range of terminal mobility in the future according to the Task Group IEEE 802.16m.

For comparison purposes, cable-based transmission systems connecting fixed subscribers in the local loop to telecommunication networks are also shown in Figure 1.3. It is clear that future wireless and mobile broadband systems are planned to reach the throughput rate of wireline systems.

Peak data rate and spectrum efficiency

There is often a confusion when comparing performance parameters of standardized air interfaces that results from not differentiating between mean values such as capacity (the maximum throughput available in a cell), throughput (the data rate perceived by a user terminal at its current location in the cell) and peak data rate (the maximum data rate available to serve a user terminal under best radio conditions).

According to its technological state-of-the-art, air interfaces are also being characterized by its spectral efficiency measured by the number of bits that can be transmitted in one Hertz bandwidth unit (bit/s/Hz). Figure 1.4 shows the performance characteristics of existing and forthcoming 'beyond the third generation' (B3G) and fourth generation (4G) mobile

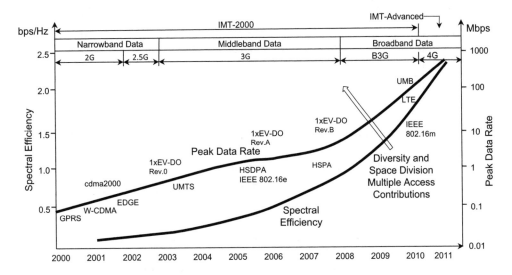

Figure 1.4 Spectral efficiency and peak data rate development over time.

system standards in terms of their peak data rate and spectral efficiency at the time of introduction.

It can be derived from Figure 1.4 that spectral efficiency almost doubles every two years, while peak data rate doubles every year. The increase of spectral efficiency over time clearly points to the fact that more data will be possible to transmit in the future in a given channel bandwidth, compared to what is possible today. This is one reason why the estimation of future spectrum bandwidth needs of mobile systems is difficult to assess. The possible contribution by multiple antenna systems to increase spectral efficiency in future (that would reduce the need for more spectrum allocation to be able to carry a predicted user traffic load) is uncertain and depends, partly, on implementation cost considerations and therefore is difficult to predict.

Roadmap of radio systems development

It is worth considering the time plans of various standardization organizations involved in the specification and further development of wireless and mobile systems and the pace of the worldwide spectrum regulation to which these activities are aligned. See Figure 1.5 for the roadmap of 3G and 4G wireless/mobile systems spanning the time interval from 2003 to 2011.

There are two standardization organizations, namely the Third Generation Partnership Project (3GPP) and Third Generation Partnership Project 2 (3GPP2) that both focus exclusively on mobile telecommunication systems. 3GPP represents the European and Asian regional standardization groups as far as they are concerned with the development of Wideband CDMA (W-CDMA). 3GPP2 is supported substantially by the TIA, an American standardization body.

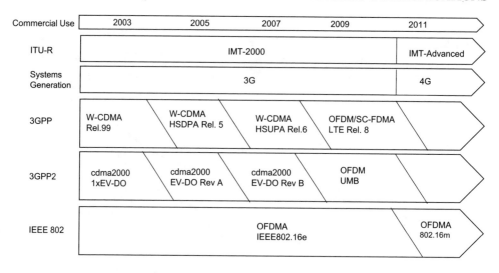

Figure 1.5 Roadmap of the evolution of wireless/mobile systems.

Both 3GPP and 3GPP2 focus in their first 3G system designs on voice and narrowband data transmission using CDMA to carry multiple connections over the same frequency channel at the same time. The main difference is the channel bandwidth applied: W-CDMA (of 3GPP) requires four times the bandwidth per channel of about 5 MHz than CDMA does, making placement of CDMA systems in the spectrum easier, if wideband (5 MHz) channels are not available. The fixed network of the radio access network is similar or even identical in architecture to both systems. With their evolution in time, both have developed towards mobile data networks for Internet access, providing application throughput rates of up to 256 bit/s as long as the serving radio cell is moderately loaded. 3GPP systems have introduced HSDPA first and have extended that technology later also to the uplink, then called HSUPA. Evolved 3GPP2 systems underline by the suffix DO (Data Only) that the respective revisions have their focus on improved data rate. It is worth noting that the CDMA component of both standards has been substantially weakened with the increased data rate, making the systems more sensitive to interference.

What can also be observed from Figure 1.5 is that the future technology for all kinds of mobile systems is Orthogonal Frequency Division Multiplexing (OFDM), a multicarrier-based transmission scheme, where multiple access to some group of carriers of the OFDM system is performed, named OFDMA.

The IEEE Working Group 802.16 originally started with an OFDM-based wireless system that soon was extended to become a mobile system in 2004, named Mobile WiMAX based on the specifications of the Task Group 16e. All tree system families, specified by 3GPP, 3GPP2 and IEEE 802, apply Frequency Division Duplex (FDD) for separating downlink and uplink transmissions in the spectrum, requiring paired channels with a substantial duplex distance. 3GPP and IEEE 802 also have specified Time Division Duplex (TDD) variants that fit better into the spectrum. Both downlink and uplink transmissions under TDD are performed on the same radio channel, separated in the time domain.

Table 1.1   Wireless/mobile systems operations philosophies.

IT community	Telecommunications community
Sell terminals (laptop, PDA, etc.), subsidize services	Sell services, subsidize mobile terminals
Certification done by WiFi/WiMAX Forum per device	Certification done per device per operator
Web 2.0 model of operator business revenue: operators rely on service providers to provide applications and get a share of the revenue. Operator revenues are limited to access	Controlled market of services and contents (see Figure 1.6). Walled garden concept. Complete control. Limited applications due to limited and closed devices
Broadband pipe	Value added service
Best efforts service	Reliable and secure service
Internet architecture. Open mobile platform	Complex and expensive infrastructure, e.g. IMS interworking
Flat-rate contract to Internet service provider with unlimited volume. Pay per use	Contract with mobile operator with volume tariffs with a trend towards flat rate component

In particular, the WiMAX WAVE releases go with TDD while 3GPP systems based on TDD have not found acceptance in the market so far.

Wireless and mobile radio systems, respectively, originate from different cultures, namely information technology (IT) and telecommunications communities. They are based on different philosophies as shown in Table 1.1.

There is a trend towards an increased fragmentation of the mobile market in terms of technologies deployed, resulting from the different economical development status of regions in the world, having different needs of mobile services that they can afford. Although it would be most efficient in terms of development costs to have one standard for all mobile systems deployed worldwide, the size of the market counted in the number of potential subscribers is so large that it can bear multiple standard systems competing. Around a hundred million terminals are being sold every year in the world, duplicating every three years. Consequently, the number of radio access technology standards will further grow.

Interworking of radio systems

Since the number of wireless and mobile systems competing in offering a service at a given location tends to ever increase, multimode terminals will be a must in the future. Such terminals should be able to hook on to the best suited air interface from a number of alternate interfaces available in a given environment. Both the traffic performance and cost shall be considered. The implementation of multimode terminals requires reconfigurability of both the transceiver hardware and protocol stack, resulting in software-defined radio terminals. Currently, terminals provide access to a number of air interfaces, each based on a dedicated transceiver and protocol suite, implemented in the mobile terminal. In future, reconfigurable mobile terminals will rely on adaptive transceivers and protocol suites, controlled by software to operate with the most suitable air interface available at a given location.

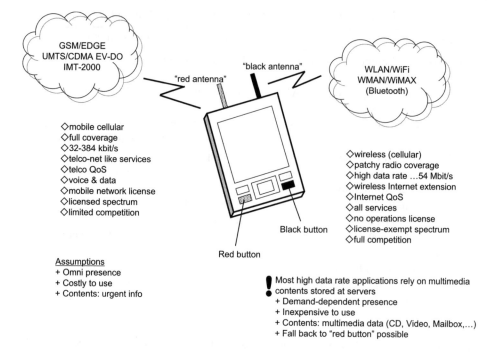

Figure 1.6 Multimode (reconfigurable) terminal able to connect to the best suited network. This figure is a slight update of Figure 18.11 of Walke (2002, p. 1037). (Reproduced by permission of © 2002 John Wiley & Sons, Ltd.)

For an understanding of the issue, Figure 1.6 shows some of the criteria to be considered when applying a multimode terminal to serve a mobile user in an optimal way (Walke 2002, Ch. 18). The characteristics of service provisioning, the class of service and the cost will be the main decision criteria for the use of one of the multiple air interfaces available from the same or a number of competing operators. Further details include:

- availability under the current conditions of terminal movement, e.g. radio coverage;

- real-time constraints of the service to be performed, for example, instantaneous when demanded and semi-instantaneous by provisioning within a given time window at a given location only;

- cost of service per time or per information unit;

- quality of service (QoS), e.g. support of real-time requirements in terms of delay and delay jitter for an interactive service, and/or application throughput rate required;

- service management, i.e. ease of use of services across different radio networks, supported by the terminal and the operators.

The two buttons (red/black) at the terminal represent internal functions used to decide which radio network to use for a given service under the current environmental conditions and costs of service.

The multitude of existing air interfaces and the increased fragmentation of radio spectrum, where a specific service is being offered, provide sufficient motivation for engaging in research towards reconfigurable terminals.

The terminals should be able to cover multiple air interfaces in a cost-efficient way. Besides this goal a demand for an increased data rate of up to 100 Mbit/s for terminals moving at high speed and that of up to 1 Gbit/s for slow-moving terminals has been identified for the future.

1.1.3 Standardization

Standardization at ITU-R

The International Telecommunication Union (ITU) is an international organization within the United Nations (UN) where governments and the private sector coordinate global telecommunication networks and services. ITU is the only body which is responsible for defining and recommending standards for international mobile telecommunication (IMT) systems. IMT-2000 encompasses all 3G mobile communication standards and their enhancements, while IMT-Advanced is the ITU name for systems beyond IMT-2000, i.e. 4G systems. ITU is responsible for the standardization of IMT systems, but the detailed standardization activities are undertaken by recognized External Organizations (EOs).

IMT-2000 is the term defined by ITU to characterize 3G mobile communication standards. Future Public Land Mobile Telecommunication System (FPLMTS) is the former name of IMT-2000. The meaning of 3G has been standardized in the IMT-2000 standardization process. A remarkable point is that the process did not standardize a technology, but it developed a set of requirements, for example, for the data rate. The Recommendation ITU-R M.1455 defines the original key characteristics of IMT-2000 radio interfaces, and represents the results of the evaluation process by the ITU Radiocommunication sector (ITU-R).

The key features of IMT-2000 are the following:

- high degree of commonality of design worldwide;

- compatibility of services within IMT-2000 and with the fixed networks;

- high service quality;

- small terminals for worldwide use;

- worldwide roaming capability;

- capability for multimedia applications, and a wide range of services and terminals.

IMT-2000 is a system with global development activity. Recommendation ITU-R M.1457 gives the detailed specifications for the radio interfaces of IMT-2000. The IMT-2000 radio interface specifications identified in Recommendation ITU-R M.1457 have been developed by the ITU in collaboration with the radio interface technology proponent organizations, global partnership projects and regional standards development organizations (SDOs).

The ITU has provided the global and overall framework and requirements, and has developed the core global specifications jointly with these organizations. The detailed standardization has been undertaken within the recognized EOs including the 3GPP and 3GPP2.

IMT-2000 comprises a single terrestrial standard which consists of two high-level groupings: CDMA, TDMA or a combination thereof. The CDMA grouping accommodates FDD direct spread, FDD multi-carrier and TDD. The TDMA grouping accommodates FDD and TDD, single carrier and multi-carrier. Recommendation ITU-R M.1457 forms the final part of the process of specifying the radio interfaces of IMT-2000 as it identifies the detailed specifications for the IMT-2000 radio interfaces. The terrestrial radio interfaces are identified as follows:

- IMT-DS (Direct Spread): W-CDMA, UTRA-FDD

- IMT-MC (Multi Carrier): CDMA2000

- IMT-TC (Time Code): UTRA-TDD, TD-SCDMA, TD-CDMA

- IMT-SC (Single Carrier): UWC-136, EDGE, GSM384

- IMT-FT (Frequency Time): DECT

See Appendix C.1 for acronyms.

Despite the promises of more feature-rich, highly interactive and high bit-rate multimedia services of 3G systems for the end users and increased revenues for the operators, the research community has perceived limitations of these systems in terms of user throughput and cost of operation. Consequently it has started to work towards B3G or 4G systems. These future systems are expected to allow subscribers to access broadband multimedia services transparently via multiple wireless and even mobile networks as if their terminals are connected via broadband cabling to the Internet. To achieve this, both the radio and fixed access network parts need substantial evolution steps to meet the expectations. The currently deployed infrastructure representing 2G and 3G technologies will need substantial evolution of existent air interfaces and the architecture of the related fixed networks that together represent a mobile radio network. Excessive investments in infrastructure and technology development are expected to make this happen, requiring time for implementation and a sufficient amount of new spectrum allocated for the operation of mobile broadband as prerequisites.

The name 'IMT-2000' denotes 3G systems, but it also encompasses the enhancements and future developments of 3G systems. ITU-R has decided on the new name for systems beyond IMT-2000 or 4G systems. Systems beyond IMT-2000 are now denoted as 'IMT-Advanced'. In the family of IMT systems, 'IMT' is the ITU accepted root name which encompasses both IMT-2000 and IMT-Advanced systems collectively. The reader is referred to Resolution ITU-R 56 for the naming of IMT systems. ITU-R has started its standardization activities by sending out a Circular Letter on an invitation to propose candidate radio interface technologies for IMT-Advanced in 2008.

The following research and development activities have been observed towards the standardization of IMT-Advanced radio interface technologies in several countries.

Japan

The mobile IT Forum (mITF) was established in 2001 for early implementation of future mobile communications systems and services such as 4G systems and mobile commerce services. It developed visions for future mobile communication systems and published several reports on 4G mobile communication systems. In 2006, the Advanced Wireless Communications Study Committee (AWCSC) was established, and replaced mITF for technical studies on 4G mobile systems. The AWCSC is responsible for:

- conducting technical studies on advanced wireless communication systems in cooperation and coordination with other related institutions in Japan and abroad;

- contributing to international standardization activities.

Korea

The Next Generation Mobile Communications (NGMC) Forum was established in 2003 to realize future mobile communications. The objectives of the NGMC Forum are to analyze technical and social trends, to establish visions on B3G, to steer advanced research and development strategies, to study spectrum use and to cooperate international organizations. The forum is developing guidelines of 4G services, 4G spectrum technologies, and it is coordinating 4G technology-related activities in Korea. The forum published vision document, white papers on system, air interface and terminals, and a document about the frequency bands for B3G with consideration on related research activities within Korea.

China

The Chinese government launched the national research project called Future Technologies for Universal Radio Environment (FuTURE) in the framework program 863 in the area of mobile communications for the time frame of the tenth 5-years-plan 2001–2005, continued in project phase 2 running until the year 2010 and aiming to achieve international leadership in mobile communications.

China-Japan-Korea B3G (CJK-B3G) is a collaboration group of China, Japan and Korea under the framework of cooperation among four Standard Development Organizations (SDOs): the Association of Radio Industries and Businesses (ARIB) and Telecommunication Technology Committee (TTC) of Japan, China Communications Standards Association (CCSA), and Telecommunications Technology Association (TTA) of Korea. In phase 1 of its activity, the CJK-B3G group discussed service requirements, service scenarios and spectrum issues, and identified technical areas and issues for B3G systems in order to form a common understanding of the B3G system. In phase 2, CJK-B3G group members exchanged information on their activities with mITF, NGMC Forum and FuTURE. They produced white papers on system requirements and enabling technologies in 2007.

Europe

In Europe, the standardization of mobile systems according to 3GPP Long Term Evolution (LTE) is supported by the Information Society Technologies (IST) research program established by the European Commission to achieve the 'Broadband for All' throughout Europe. Since the Fifth Framework Programme (FP5) (1998–2002) of the European Union (EU) Commission, research activities have been launched on systems beyond 3G. The Wireless World Initiative New Radio (WINNER) project is one example of the Sixth Framework Programme (FP6) (2002–2006) that is working towards the new radio interface for systems beyond 3G. The ETSI is taking part in the specification of 3G systems evolution.

United States

The IEEE, a globally operating organization with its headquarters in the USA, has developed standards for WLANs and has expanded that work to drive the development of future systems with enhanced capabilities such as interworking, meshing of base stations and support of mobility management that will have substantial impact on 3G and 4G systems development. Another major player in North America is the Defense Advanced Research Projects Agency (DARPA). The DARPA neXt Generation (XG) communications program is developing a technology to allow multiple users to share the spectrum through adaptive mechanisms.

1.2   Trends in Spectrum Usage

This section gives a background to those who are not familiar with radio spectrum and the concept of spectrum allocation currently used for spectrum management. It also mentions a flexible spectrum use for the future.

1.2.1   Physical properties of radio spectra

Radio spectrum is the radio frequency portion of the electromagnetic spectrum. The radio spectrum is a valuable and limited natural resource as it is generally available within a range of between 3 kHz and 3000 GHz for communications purposes. This range has been split into radio spectrum bands, e.g. high frequency (HF), very high frequency (VHF) and ultra high frequency (UHF) bands. Only frequency bands between 9 kHz and 275 GHz are currently allocated to radio communication applications by ITU-R on a global basis.

Figure 1.7 shows radio spectrum bands, their corresponding wavelengths/frequencies and example radio applications. Different portions of the radio spectrum have different physical properties due to the different wavelength of the radio waves. These physical properties include, among others, path loss, propagation mechanisms and absorption.

Path loss

Line-of-sight (LOS) is the direct propagation of radio waves between transmitter and receiver antennas. The radio waves travel in straight lines just like light waves travel. In the free space, all radio waves obey the inverse square law where the power density of a radio wave is proportional to the inverse of the square of the distance d between transmitter and receiver antennas: d^{-2} propagation. Mobile cellular systems use frequencies at which radio waves primarily propagate in the non-LOS mode. Then the d^{-4} propagation is observed.

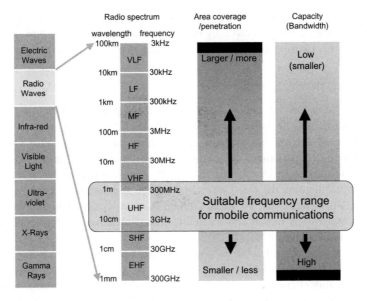

Figure 1.7 Frequency range suitable for mobile communication.

An empirical formula for the dependency of path loss L_p on the frequency f can be generally expressed as (COST 231 1991; Hata 1980; Walfisch and Bertoni 1988)

$$L_p = A_d \log(d) + A_f \log(f) + \text{constant}, \tag{1.1}$$

where d and f are the distance between mobile and base stations, and carrier frequency, respectively. A_d and A_f are coefficients which have distance and frequency dependencies, respectively. As the name says, the path loss describes the deterioration of the radio signal power on its way from a transmitter to the receiver terminal. In urban areas, $A_d = 38$ and $A_f = 20$ could be obtained from measuring campaign (Okumura *et al.* 1968). It can be concluded that the path loss increases proportionally to around the second power of the carrier frequency in urban areas.

Propagation mechanisms

Radio propagation has the following three key elements: diffraction, reflection and scattering. When a radio wave is obstructed by an irregular surface, secondary waves are generated. They bend around and reach behind the obstacles. Reflection occurs when a radio wave encounters obstacles larger than its wavelength. Scattering happens when a radio wave travels through a medium that contains many small objects compared to its wavelength.

Diffraction depends on the wavelength of the radio wave and the size of the obstacles. Due to the nature of the waves, radio waves at a lower frequency diffract more easily around large obstacles such as hills and smooth mountains. Diffraction is very important for the mobile systems to cover the service area with sufficiently strong signal energy. Radio waves diffracted by small obstacles such as buildings at high frequency around UHF and SHF (see Figure 1.7) can travel over roof edges of the buildings into the streets.

The diffracted waves and their reflected waves overlay each other resulting in multi-path propagation. They are finally received by the antenna of a mobile radio terminal. Non-LOS propagation is more easily obtained at a lower frequency range, say VHF.

Absorption

The radio wave at a low frequency travels easily through walls, bricks and stones. Easy penetration into houses or buildings in the UHF mobile cellular system improves the indoor coverage of the mobile system. As the frequency of the radio wave becomes higher, absorption becomes more dominating. In SHF or at higher frequencies, absorption by molecular resonance in the atmosphere, such as gas, water, vapor and oxygen, is a major factor in attenuating the radio wave energy. The attenuation by heavy rain and snow sometimes results in outage of the radio communications at the SHF frequency band and above. This is particularly important for satellite communications in a higher portion of SHF bands, e.g. around 12 GHz.

Taking the above properties into account, different portions of the radio spectrum are better suited to different purposes.

A radio spectrum channel has a bandwidth giving the width of the range of frequencies on which an information signal travels. A broader bandwidth is required in order to convey higher data rate information. This suggests that the higher frequency bands in the spectrum are more suitable for broadband communications.

However, terrestrial mobile radio systems depend mainly on non-LOS radio communications exploiting radio propagation mechanisms such as diffraction, reflection and scattering. Diffracted radio waves at frequencies above 5 GHz are significantly attenuated, resulting in a high path loss making non-LOS communications difficult to realize. It is therefore concluded that the suitable frequency range for mobile systems is below 5 GHz.

1.2.2   Spectrum allocation and identification

Allocation

Each country has an administration that sovereignly regulates and manages spectrum use within its territory. In order to manage radio spectrum appropriately, various kinds of radio stations are categorized into radio services such as mobile service, fixed service, broadcasting service and satellite service. The categorization is based on operational and administrative differences rather than those in technical characteristics. The radio spectrum is also divided into many portions of the frequency bands. Administrations allocate one or more of the services to these discrete portions of the bands in order to achieve an efficient use of spectrum and to avoid harmful interferences among different radio services. Each administration develops its own national table of frequency allocations (see Section 2.2) defining which portion of the spectrum shall be used for what kind of a particular radio service. According to this national allocation table, the administration manages spectrum in its territory as a basis for an orderly and efficient use of the radio spectrum.

Since radio waves propagate across national borders, the administrations cannot manage the radio spectrum separately and independently. International cooperation in spectrum management is required. The international cooperation can be achieved on a bilateral or multi-lateral basis. Regional organizations such as the European Conference of Postal and

Telecommunications Administrations (CEPT), the Asia-Pacific Telecommunity (APT), and the Inter-American Telecommunication Commission (CITEL) have been set up for this purpose. ITU is a global organization that deals with the international cooperation of the spectrum management. The administrations cooperate in spectrum management through the ITU. ITU member countries must obey the agreements that are reached at ITU, giving them a treaty status. Among the agreements, the Radio Regulations (RR) provide basic principles and terminologies which are commonly used by the administrations. The RR also provides an international frequency allocation table which has been agreed at the World Radiocommunication Conferences (WRCs). The RR has been updated at WRCs. National frequency allocation tables are harmonized with the international allocation table in the RR. It should be noted that a discrete portion of the spectrum is 'allocated' to one or more of the 'services' without referring to specific radio systems. The administrations can assign a portion of the spectrum to any kind of radio system that belongs to the service to which the portion of the spectrum is allocated. Since cellular systems emerged quite recently, the spectrum that has already been allocated to the mobile service is small compared to the spectra of other services such as the broadcasting and satellite services.

ITU-R arranges the WRC every three to four years to review and revise the RR, which forms the international treaty governing the use of the radio frequency spectrum and satellite orbits. WRC is therefore the only authority to decide on frequency allocations. For an issue to be considered at the WRC, it needs to be included on the WRC agenda, which is developed at the preceding conference. Proposals for an item to be included on the WRC agenda need to be discussed before the preceding conference. A conference preparatory meeting (CPM) is organized before each WRC to prepare a consolidated report to support the work of the WRC.

Due to the above procedures, it takes more than 3 to 4 years from the beginning of the discussions until the allocation itself. From the allocation of frequency bands for some service, it usually takes several years until the actual frequency bands can be used for the service. Therefore, frequency-related matters must be considered well in advance before the actual need for the bands.

Identification

In mobile communication systems such as IMT-2000, there was a further need to specify what kind of particular radio system uses which portion of the spectrum band to ensure global roaming of radio terminals that belong to the same radio system and use the same spectrum band worldwide. The concept of 'spectrum identification' was created for this purpose. Since the spectrum identification is further made in addition to the spectrum allocation, in general the portion of the band to be identified should have a primary allocation before the identification. Global identification is particularly important for mobile systems in that it ensures that the same radio systems can deploy in the same frequency band on a global basis. This results in economies of scale (cost reduction by mass production) and global roaming of mobile terminals.

1.2.3 Flexible use of spectrum

Concepts of fixed and dynamic spectrum allocation are discussed in the following.

At present, the interference between radio systems operated by different operators is kept below some margin through a fixed allocation of frequency bands to the operators, with sufficient guard bands in between. Thereby, the mobile radio spectrum is segmented and some bands of the spectrum are exclusively licensed by a regulator to one operator, while other bands are assigned to other operators. This is seen as a precondition for competition between operators in the mobile services market, bringing down the cost of the use of mobile services experienced by subscribers. A radio band licensed to an operator is, typically, subdivided into multiple radio channels, and an operator's radio network is planned in a way such that the same channel is being spatially reused at a distance large enough so that the path loss of co-channel signals is high enough to avoid substantial signal interference. This model of spectrum allocation to mobile operators has been applied successfully over decades.

The spectrum range suitable for mobile radio services, see Figure 1.7, is allocated to only a small percentage to mobile services. Many other services such as radio broadcast, television (TV) broadcast, fixed and mobile satellite, fixed radio links and radio navigation have successfully competed for allocation in earlier days. In addition, public and military services have their exclusive allocations for similar services. It is easy to check by measurement that only very small parts of the range 'suitable for mobile use' are really being used by the respective licensees. Measurement campaigns at most locations, which graphically represent the signal energy received over 24 hours per spectrum unit in the range of 300 MHz to 3 GHz, typically, show a blue carpet (where blue color represents 'no signal received') with some narrow frequency band lines covering all the time, representing local radio stations and active mobile operators. From that it can be concluded that spectrum is not a scarce resource, but that spectrum is allocated mostly to licensees that, typically, rarely operate the respective service or operate the service only limited to some small spatial area, or both. This observation is not new and has generated activities for more efficient use of the radio spectrum by introducing flexibility of use.

TV broadcast channels occupy a large portion of the spectrum best suited for mobile services. Since operators of TV channels are known to operate their channels only in metropolitan areas and to keep many channels elsewhere unused, the IEEE 802.22 Working Group on wireless regional area networks (WRANs) is to develop a standard for a cognitive radio-based PHY/MAC air interface for use by license-exempt devices on a non-interfering basis in the spectrum allocated to the TV Broadcast Service. Besides this, the IEEE 802.19 Coexistence Technical Advisory Group (TAG) will develop and maintain policies defining the responsibilities of 802 standards developers to address issues of coexistence with existing standards and other standards under development. Worth noting is also the activity of IEEE 1900 Working Group developing supporting standards dealing with new technologies and techniques being developed for next generation radio and advanced spectrum management. This activity is known as Dynamic Spectrum Access Networks (DySPAN).

Experiences exist for the dynamic spectrum use of systems following the same or different air interface standards operated in license-exempt bands, known as Industrial, Scientific and Medical (ISM) bands. The operation of IEEE 802.11 WLAN is a predominant example of this. The problem with shared use of a frequency channel by non-coordinated systems is that the information transmission range over radio between a transmitter and a receiver substantially differs from the interference range, within which any other receivers might harmfully be affected. Therefore it can happen that a TV broadcast receiver would catch much 'noise' (degradation of the signal quality) from interference by a terminal operated in the same channel according to the standard IEEE 802.22.

Frequency sharing rules (FSRs) have been suggested to avoid unwanted systems interference (Motorola 1994). FSRs have been evaluated by modeling the statistical interference to predict the probability of non-acceptable mutual interference for systems such as UMTS versus DECT, GSM versus TErrestrial Trunked RAdio (TETRA) (Lott and Scheibenbogen 1997). The focus there is on the prediction of the QoS of mobile services operated adjacently in the spectrum, with too small a guard band in between.

Present spectrum allocation has some shortcomings, especially when considering efficient use of a given radio band as follows:

- Almost all radio systems experience time-dependent load characteristics. Most services have a certain, predictable load pattern over the course of a day, with peak times and low-load periods. Currently, spectrum is assigned for serving the peak load. However, most of the spectrum will be unused for long periods of time. Dynamic spectrum allocation might take advantage of this by allowing systems to breath in temporal spectrum occupancy according to its current load under given rules.

- Spectrum efficiency can be increased by choosing the optimal transmission technology for a given service. Digital contents can be transmitted via both a point-to-point UMTS link and a broadcast DAB link. Whenever several mobile terminals are requesting the same contents, e.g. 'Wireless News', a broadcast service may be more efficient compared to point-to-point transmission. Thereby, bandwidth to transmit multiple copies of the same data may be saved.

- Intelligent systems capable of detecting or predicting interference may be able to avoid mutual interference. Thus, spectrum reserved for guard bands may be used for data transmission.

- Current mobile systems cannot mutually exchange information on their respective spectrum usage. This might be overcome in favor of more cooperation between systems, resulting in a more efficient use of unused spectrum.

Cognitive radio is the key to more spectrum-efficient use of the radio spectrum. It will allow dynamic spectrum use according to the current needs of a system, either under central control, e.g. by following the information broadcast by a pilot channel, or under noncentral control based on game theoretic considerations or similar (Walke *et al.* 2006).

1.3 Spectrum Allocation: Why and How

This section describes why spectrum requirement estimation is required for spectrum allocation and how the estimation is conducted.

1.3.1 Requirement estimation for allocation

The users' demands for broadband mobile and wireless communications are increasing as they can provide a wide variety of attractive applications/services. In order to satisfy the users' demands, new radio interface technologies for IMT-Advanced will be required after the year 2010 that support the data rate of 100 Mbit/s in full mobility environments and 1 Gbit/s in nomadic environments. These radio interface technologies require continuous and broader spectra.

Since the current spectrum usage is determined based on an allocation table, the allocation table in the RR should be revised at the ITU-R WRC. In order to revise the allocation table, as a justification for the application for allocation of new spectra, the spectrum required for future services and systems must be predicted as accurately as possible, taking into due consideration future market trends including applications and services as well as future technology trends.

1.3.2   Method of estimation

To conduct estimation of spectra to be used in addition to the spectra allocated already for mobile services during the years 2010 to 2020, ITU-R set the following two principles.

- A spectrum estimation methodology should appropriately handle the traffic arising from a wide variety of future mobile applications and services.

- It should also consider multiple delivery mechanisms to accommodate the traffic.

Based on these principles, ITU-R extensively collected and analyzed the data on applications and services as well as market and technology trends in the year 2010 onwards.

As for applications and services, ITU-R investigated a wide variety of services and their similarities, and finally classified these services into 20 service categories (SCs). Moreover, it defined six service environments (SEs) by using service usage patterns and teledensities (density of terminals using a specific telecommunications service) in order to analyze the users' behavior of service usage. Regarding radio access systems, ITU-R defined four Radio Environments (REs) by considering cell deployment scenarios and four Radio Access Technique Groups (RATGs) by considering their system characteristics and their deployments; see Section 4.2.

In ITU-R, the spectrum requirement has been calculated, focusing on a specific radio system, with radio parameters of the system and services associated with it. ITU-R has taken a new approach in which both all possible wireless applications and all possible mobile radio systems that can accommodate them are first considered, and then has systematically categorized them into a limited number of service categories and RATGs for spectrum requirement calculation.

The new ITU-R methodology employs a system capacity calculation algorithm for packet-switched service categories in addition to that for the conventional circuit-switched service categories. The algorithm attains a traffic multiplexing gain, which avoids an overestimation of the spectrum bandwidth. This new methodology is presented in Chapter 4 of this book.

2

Utilization of Radio Frequencies

Hitoshi Yoshino, Naoto Matoba, Pekka Ojanen and Bernhard H. Walke

In order to use the radio spectrum efficiently, the characteristics of the radio spectrum and the demand for the spectrum should be considered. The characteristic of the spectrum varies according to the wavelength of the radio waves. In this chapter, we review the utilization of radio waves.

Section 2.1 overviews the spectrum usage based on the characteristics of the spectrum. Spectrum demands have continuously developed according to the introduction of an increased number of radio-based services. Accordingly, ITU-R has periodically reviewed the spectrum allocation based on the demand at each time. Section 2.2 describes how ITU-R manages the radio spectrum. ITU-R has introduced a concept of *radio services* involving the transmission, emission and/or reception of radio waves for specific telecommunication purposes. ITU-R also considers radio services when it allocates spectrum to them in order to achieve efficient and economic use of the spectrum. In Section 2.3, the radio services defined by ITU-R are described. In Section 2.4, some recent radio communication systems are presented together with their characteristics. In particular, cellular mobile communication systems are explained in detail.

2.1 Spectrum Usage Overview

The *radio spectrum* covers the frequency range of electromagnetic waves between 3 kHz and 3 THz (= 3000 GHz). The radio spectrum is divided into nine frequency bands as shown in Table 2.1. The spectrum usage is primarily determined by its propagation mode and availability of bandwidth for radio applications. It is summarized in Figure 2.1.

Spectrum Requirement Planning in Wireless Communications   Edited by Hideaki Takagi and Bernhard H. Walke
© 2008 John Wiley & Sons, Ltd

Table 2.1   Frequency bands.

Band	Acronym	Frequency	Wavelength
Very low frequency	VLF	3 kHz–30 kHz	100–10 km
Low frequency	LF	30 kHz–300 kHz	10–1 km
Medium frequency	MF	300 kHz–3 MHz	1000–100 m
High frequency	HF	3 MHz–30 MHz	100–10 m
Very high frequency	VHF	30 MHz–300 MHz	10–1 m
Ultra high frequency	UHF	300 MHz–3 GHz	100–10 cm
Super high frequency	SHF	3 GHz–30 GHz	10–1 cm
Extremely high frequency	EHF	30 GHz–300 GHz	10–1 mm
Sub-millimeter wave		300 GHz–3 THz	1–0.1 mm

Band	Propagation mode	Radio applications
VLF 3 kHz – 30 kHz	- Guided between the earth surface and the ionosphere	- Submarine communications - Radio navigation systems (OMEGA)
LF 30 kHz – 300 kHz	- Guided between the earth surface and the ionosphere - Ground waves	- AM radio broadcast, - Telegraph communication, - Radio navigation (LORAN) and aircraft beacon - Time signal
MF 300 kHz – 3 MHz	- Ground waves - Sky waves	- AM radio broadcast
HF 3 MHz – 30 MHz	- Sky waves	- Long range terrestrial communication system - International radio broadcast
VHF 30 MHz – 300 MHz	- Direct wave (usually) - Tropospheric ducting - Sky waves (during high sunspot activity)	- FM radio and TV broadcast - Terrestrial navigation e.g. VOR for aircraft - Aircraft and maritime communications,
UHF 300 MHz – 3 GHz	- Direct wave	- FM radio and TV broadcast - Cellular phone systems - TErrestrial Trunked RAdio (TETRA) systems - Global Positioning System (GPS)
SHF 3 GHz – 30 GHz	- Direct wave	- Microwave fixed radio links - Wireless LAN - Modern shipborne and meteorological radars - Fixed satellite systems
EHF 30 GHz – 300 GHz	- Direct wave limited by absorption	- High-speed microwave data links - Radio astronomy and remote sensing - Radar systems with very high resolution - Intersatellite communication links
300 GHz – 3 THz	- Direct wave limited by absorption	- Medical imaging - High latitude communications (aircraft – satellite)

Figure 2.1 Radio spectrum, their primary propagation mode and radio applications.

2.1.1   VLF band

The very low frequency (VLF) band overlaps the audio frequency spectrum (20–20 000 Hz). The VLF band radio waves are transmitted by electromagnetic energy while the audio signals (sounds) are transmitted by atmospheric compression and expansion. The radio waves on the VLF band propagate through the waveguide formed between the earth surface and ionosphere at long distances. Its attenuation ranges between 0.1 and 0.5 decibel (dB) per 1000 km.[1] The VLF waves also penetrate water below 10 to 40 m, depending on its frequency and the salinity of the water. The longer the radio waves are, the deeper they penetrate the ocean. VLF waves are therefore used for submarine communications near the water surface. It is difficult to use this band for telephony since the available bandwidth is very small. Telecommunication in this band is therefore based on telegraph signals transmission operating at very low data rate. The VLF waves are also used for radio navigation using very simple signals that occupy a very small portion of the band. One example is the OMEGA Navigation System (operated from 1971 to 1997), the first worldwide navigation system using hyperbolic radio navigation techniques used by aircrafts, ships and submarines.

2.1.2 LF band

The low frequency (LF) band radio waves can be transmitted over very long distances along the ground surface. The propagation mode in which radio waves propagate close to the Earth's surface is called ground-wave propagation or surface-wave propagation. The ground waves travel further beyond the horizon since the Earth's surface pulls the radio waves and keeps them along the ground. Until the 1930s, the band was used mainly for telegraph communications. Since a radio station using the LF band needs a large antenna and high transmission power, the telegraph communication in this band decreased after the HF band radio communication systems were developed. One part of this band is used for Amplitude Modulation (AM)-based radio broadcast services in Europe, parts of North America, and Asia. Since the radio propagation in this band is only by ground wave, the radio waves are not affected by varying propagation paths between the transmitter and receiver, and ionosphere. The LF band radio waves are therefore used for aircraft beacon, navigation, e.g. LOng RAnge Navigation (LORAN) system, and radio clock signal stations that provide high-precision time signals.

2.1.3 MF band

The medium frequency (MF) band radio waves radiated from the ground are reflected back to the Earth at a low level in the ionosphere, the D region by day and the E region by night, which are located 50–90 km and 90–140 km above the ground, respectively. The propagation of radio waves refracted back to the Earth's surface by the ionosphere is called sky-wave propagation. Since the sky-wave propagation in this band is stable over a long distance, parts of the band are used for AM-based broadcast services. The broadcast radio stations operated in this band need large antennas and high power transmitters but the receiver devices can be simple and small.

---

[1]The decibel (dB) is a logarithmic unit for measuring the signal power X at the receiver relative to a reference level X_0 at the transmitter by

$$10 \log_{10}\left(\frac{X}{X_0}\right).$$

Thus -0.1 dB means $X = 10^{-0.1/10}X_0 = 0.977X_0$ and -0.5 dB means $X = 10^{-0.5/10}X_0 = 0.891X_0$. Hence the VLF band radio waves do not attenuate over a long distance.

2.1.4   HF band

The high frequency (HF) band radio waves also propagate in the sky waves, being reflected by the ionosphere and travel all over the world by multihop propagation between the ground and the ionosphere. The HF band is used for medium and long-range terrestrial communications and international radio broadcast services. As for reflection by the ionosphere, there are upper and lower frequency bounds for radio communications. If the frequency is above the upper bound, the radio waves will pass straight through the ionosphere. If the frequency is below the lower bound, the radio waves are considerably attenuated due to the absorption in the D region of the ionosphere. The range of usable frequencies for HF communications varies:

- throughout the day

- with the seasons

- with the solar cycles

- with solar activity

- from place to place

because the ionization of the ionosphere depends on the solar radiation. In HF radio communications and broadcasting, radio stations select the carrier frequency for operation, dynamically, in the HF band to take the above-mentioned factors into account.

2.1.5   VHF band

The very high frequency (VHF) band radio waves are usually not reflected by the ionosphere, with some exceptions below 150 MHz. VHF telecommunication usually applies direct-wave propagation, where the radio waves travel from one antenna to another in a straight line called Line-Of-Sight (LOS). Since the radio waves cannot travel over the horizon, its transmission range is limited to the local area. The VHF band is less affected by atmospheric noise and interference from electric devices than HF and lower frequency bands. Radio wave propagation in the VHF band is less affected by buildings and other obstacles than waves at higher frequency bands because the VHF radio waves can reach behind the obstacles due to their diffraction. VHF waves are more easily blocked by terrain than waves at HF and lower bands. Therefore, the VHF band is suitable for short-distance terrestrial communications, e.g. land mobile, aircraft and maritime communications, with a distance range up to LOS, e.g. 20 to 30 km. This band is also heavily used for Frequency Modulation (FM) radio and television (TV) broadcasting services. It is further used for terrestrial navigation systems, e.g. VHF Omnidirectional radio Range (VOR) for aircraft navigation.

The VHF band also has two unusual propagation modes: tropospheric ducting and sporadic E region propagation. Tropospheric ducting occurs in front of and parallel to an advancing cold weather front, especially under the condition that there is a difference in humidity between the cold and warm air masses. The VHF band radio waves can propagate inside the duct for hundreds of kilometers. Sporadic E region propagation occurs when, occasionally, ionized atmospheric gas is formed in the ionosphere E region due to the high solar activity. The sporadic E region can reflect not only the HF band radio waves but also the radio waves of the lower part of the VHF band that usually go through the ionosphere.

These unusual propagation modes occasionally allow long-distance communications at VHF band and sometimes cause harmful interference to distant radio stations.

2.1.6 UHF band

Radio communications in the ultra high frequency (UHF) band also use the direct-wave propagation mode. UHF band signals are less affected by reflections in the ionosphere than signals in VHF and lower frequency bands, resulting in less harmful interference from a distant place. In addition, the UHF band signals are more attenuated by the atmospheric moisture than signals in the VHF band, since the attenuation increases as the frequency of the radio wave becomes higher. This is because radio waves are partially absorbed by atmospheric moisture, which attenuates the radio signal strength during propagation. The UHF band is therefore more suitable for shorter-distance radio communications, say 10 to 15 km. The UHF band is used for TV broadcast services. It is also used for two-way communication systems, including cellular radio systems and TErrestrial Trunked RAdio (TETRA) systems, known as walkie-talkies. The band is also used for fixed wireless links and medium-range radars.

The UHF band has a relatively short wavelength compared to VHF or lower bands. This leads to smaller sizes of antenna and transceiver device.

Since the UHF radio waves can go through the ionosphere, the satellite communication systems also use this band, including the Global Positioning System (GPS) and mobile satellite systems.

2.1.7 SHF band

Radio signals propagate in the super high frequency (SHF) band according to the direct wave mode. Due to the availability of wider bandwidth at SHF compared to UHF and lower bands, radio applications that require higher bandwidth use this band. Future mobile broadband systems such as IMT-Advanced are therefore expected to operate in the SHF band. However, at the SHF band, radio waves suffer from higher attenuation than in lower bands.

This property has been exploited for radio applications that require a range of coverage in the order of some 100 m up to some km, such as microwave fixed radio links, Wireless Local Area Networks (WLAN), most modern radars (shipborne radars and meteorological radars), fixed satellite service (FSS) systems, and wireless Universal Serial Bus (USB) devices using Ultra Wide Band (UWB) transmission.

2.1.8 EHF band

The extremely high frequency (EHF) band substantially suffers from atmospheric attenuation, severely limiting long distance communications. This, however, allows for more efficient frequency reuse than possible at lower frequency bands. The small wavelength also contributes to downsize the antenna and transceiver devices. This band is used for high-speed microwave data links in some countries but has not yet been used extensively due to the limited communication range. Currently, the band is commonly used in radio astronomy and remote sensing. The band is also used for radar systems with very high resolution and intersatellite communication links, where no atmospheric attenuation is present.

2.2   Spectrum Management by ITU

Early days of ITU

The first radio transmission systems in the 1890s had a very wide bandwidth as spark-gap transmitters were applied. In 1903 closed resonance circuits were invented, enabling the radio signal to be adjusted to a given carrier frequency. It became possible to filter an incoming signal out of a band used for multiple transmissions in parallel. Communication channels were relatively narrow band (below 50 kHz bandwidth) at carrier frequencies below 1 MHz. Besides military applications, shipping companies were the early adopters of mobile radio communication, as they wanted to know where their ships were roaming on the ocean. Until today, over the decades the acceptance of radio-based services expanded with exponential growth occupying a wide range of the spectrum. In line with this, the service bandwidth of communication links evolved from narrow band 12.5 kHz for voice up to 20 or even 40 MHz currently discussed for mobile broadband communication channels in the fourth generation (4G) radio networks. Radio transmissions have evolved from single transmission links to a range from simple links to complex networks. Today we are surrounded by manmade radio signals.

   Initially the need for frequency coordination was not very urgent. However, as the radio transmissions evolved to cover wide distances, the need for national and international coordination was soon recognized. The International Telegraph Union (ITU) was established in 1865 to take care of telegraph-related regulation. As a consequence of the invention of the wireless telegraphy, the first International Radiotelegraph Conference was held in 1906. The Conference signed the first International Radiotelegraph Convention. The annex to the Convention contained the first regulations governing wireless radiotelegraphy. An early history of ITU can be found on the web site 'http://www.itu.int/aboutitu/overview/history.html'.

Radio regulations by ITU-R

Today the usage of radio frequencies is administered globally by the *International Telecommunication Union* (ITU) that evolved through various steps from the original ITU. The *Radio Regulations* (RR) of the Radiocommunication Sector of the ITU (ITU-R), more or less originating from the Convention signed in 1906, is the global basis for the regulation of the spectrum use. The RR is a binding intergovernmental treaty governing the use of radio spectrum. ITU-R divides the world into three Regions as shown in Figure 2.2. A precise definition of the boundaries between Regions may be found in Article 5 of the ITU Radio Regulations. There are differences in the allowed spectrum use among the Regions.

Radio services

The main tool for the ITU-R in defining the allowed spectrum use is the so-called *spectrum allocations*. Portions of the spectrum, *spectrum bands*, are allocated to different *Radio Services* (Services). The whole radio spectrum between 9 kHz and 400 GHz is allocated in the RR to different Services. Each Service represents a certain kind of usage, which is typically characterized by transmission characteristics or by the state of the transmitters, e.g. related to location or mobility. The ITU-R Services do not define 'services' as categories of contents.

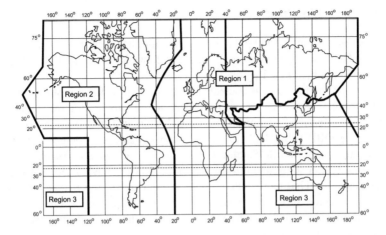

Figure 2.2 Three Regions of the world divided by ITU-R for frequency allocation.

They are, rather, categorization based on the kind of transmitters. There are currently around 40 different ITU Services defined, including, for example, the following:

- Mobile service (MS)

- Broadcast service (BS)

- Fixed service (FS)

- Fixed satellite service (FSS)

- Mobile satellite service (MSS).

These are discussed in detail in Section 2.3.

Radio spectrum

The whole radio spectrum is allocated to the different ITU Services. There is no unused available spectrum. However, as an exception there are 21 defined bands where all emissions are prohibited. The bands include 1400–1427 MHz, 2690–2700 MHz and 19 bands above 10 GHz. Such bands are used for reception only.

Visual examples of the full radio spectrum use in some countries can be found on the following web sites:

- Japan

  http://www.tele.soumu.go.jp/e/search/myuse/use0303/index.htm

- Finland

  http://www.ficora.fi/englanti/document/Use_of_radio_spectrum.PDF

- United States

  http://www.ntia.doc.gov/osmhome/allochrt.pdf

Table of frequency allocations

Today the RR contains over 1000 pages of information that describe how spectrum may be used globally. The *Table of Frequency Allocations* shows all allocations covering the whole radio spectrum in all three ITU Regions.

An example of the format and contents of the Allocation Table is shown in Figure 2.3. In this example some bands have the same allocation globally in all three Regions, whereas other bands have different allocations in one or each Region. The allocations shown in upper-case letters (e.g. MOBILE-SATELLITE) are *primary allocations*, whereas allocations shown in lowercase letters (e.g. Mobile-Satellite) are *secondary allocations*. Services with primary allocation have a higher status and priorities over the services having secondary allocation.

Typically, each band has more than one allocation, which means that national administrations may choose which Service the band is to be used for. In practice it makes sense to take into account how the band is used elsewhere. In some other cases it may mean that several Services can actually share the band.

Footnotes

In addition to Allocations there are *footnotes* related to each band. Some example footnotes taken from the RR are shown in Figure 2.4. Footnote 5.387 is an example of national exceptions compared to the regional information given in the Table of Frequency Allocations. Footnote 5.388 is exceptional among the footnotes as it *identifies* certain bands out of those that have the allocation for Mobile Service for a certain technology, namely IMT-2000. IMT-2000 is actually a family of technologies and not a single technology, but still ITU-R usually does not define the spectrum usage more than to the allocation level. Footnote 5.338A is an example of regional information given additionally to what is in the Table of Frequency Allocations.

National implementation

The ITU Table of Frequency Allocations defines only a framework for the national spectrum use and in practice there is considerable freedom for national administrations actually to make spectrum bands available for particular use.

Let us take the terrestrial cellular mobile systems as an example. From the RR point of view, such systems could be deployed in any bands having an allocation for Mobile Service. The RR has many bands that have either a primary or secondary allocation for Mobile Service, therefore different countries could pick very different combinations from them resulting in different bands to be used in different countries.

For analog cellular systems the best option out of all bands that were suitable from the allocation point of view was to choose those that were already widely used by the cellular systems elsewhere. That approach allowed availability of equipment, possibility for international roaming and economies of scale. However, the result was still globally very fragmented.

1710–2170 MHz

Allocation to services		
Region 1	**Region 2**	**Region 3**
1710–1930	FIXED MOBILE 5.380 5.384A 5.388A 5.388B 5.149 5.341 5.385 5.386 5.387 5.388	
1930–1970 FIXED MOBILE 5.388A 5.388B 5.388	**1930–1970** FIXED MOBILE 5.388A 5.388B Mobile-Satellite (Earth-to-space) 5.388	**1930–1970** FIXED MOBILE 5.388A 5.388B 5.388
1970–1980	FIXED MOBILE 5.388A 5.388B 5.388	
1980–2010	FIXED MOBILE MOBILE-SATELLITE (Earth-to-space) 5.351A 5.388 5.389A 5.389B 5.389F	
2010–2025 FIXED MOBILE 5.388A 5.388B 5.388	**2010–2025** FIXED MOBILE MOBILE-SATELLITE (Earth-to-space) 5.388 5.389C 5.389E 5.390	**2010–2025** FIXED MOBILE 5.388A 5.388B 5.388
2025–2110	SPACE OPERATION (Earth-to-space) (space-to-space) EARTH EXPOLATION-SATELLITE (Earth-to-space) (space-to-space) FIXED MOBILE 5.391 SPACE RESEARCH (Earth-to-space) (space-to-space) 5.392	
2110–2120	FIXED MOBILE 5.388A 5.388B SPACE RESEARCH (deep space)(Earth-to-space) 5.388	
2120–2160 FIXED MOBILE 5.388A 5.388B 5.388	**2120–2160** FIXED MOBILE 5.388A 5.388B Mobile-Satellite (space-to-Earth) 5.388	**2120–2160** FIXED MOBILE 5.388A 5.388B 5.388
2160–2170 FIXED MOBILE 5.388A 5.388B 5.388 5.392A	**2160–2170** FIXED MOBILE MOBILE-SATELLITE (space-to-Earth) 5.388 5.389C 5.389E 5.390	**2160–2170** FIXED MOBILE 5.388A 5.388B 5.388

Figure 2.3 Example of the Table of Frequency Allocations in the ITU Radio Regulations.

5.384A The bands, or portions of the bands, 1710–1885 MHz and 2500–2690 MHz, are identified for use by administrations wishing to implement International Mobile Telecommunications-2000 (IMT-2000) in accordance with Resolution **223 (WRC-2000)**. This identification does not preclude the use of these bands by any application of the services to which they are allocated and does not establish priority in the Radio Regulations. (WRC-2000)

5.385 *Additional allocation*: the band 1718.8–1722.2 MHz is also allocated to the radio astronomy service on a secondary basis for spectral line observations. (WRC-2000)

5.386 *Additional allocation*: the band 1750–1850 MHz is also allocated to the space operation (Earth-to-space) and space research (Earth-to-space) services in Region 2, in Australia, Guam, India, Indonesia and Japan on a primary basis, subject to agreement obtained under No. 9.21, having particular regard to troposcatter systems. (WRC-03)

5.387 *Additional allocation*: in Azerbaijan, Belarus, Georgia, Kazakhstan, Mongolia, Kyrgyzstan, Slovakia, Romania, Tajikistan and Turkmenistan, the band 1770–1790 MHz is also allocated to the meteorological-satellite service on a primary basis, subject to agreement obtained under No. **9.21**. (WRC-03)

5.388 The bands 1885–2025 MHz and 2110–2200 MHz are intended for use, on a worldwide basis, by administrations wishing to implement International Mobile Telecommunications-2000 (IMT-2000). Such use does not preclude the use of these bands by other services to which they are allocated. The bands should be made available for IMT-2000 in accordance with Resolution **212 (Rev. WRC-97)**. (See also Resolution **223 (WRC-2000)**.) (WRC-2000)

5.388A In Regions 1 and 3, the bands 1885–1980 MHz, 2010–2025 MHz and 2110–2170 MHz and, in Region 2, the bands 1885–1980 MHz and 2110–2160 MHz may be used by high-altitude platform stations as base stations to provide International Mobile Telecommunications-2000 (IMT-2000), in accordance with Resolution **221 (Rev. WRC-03)**. Their use by IMT-2000 applications using high-altitude platform stations as base stations does not preclude the use of these bands by any station in the services to which they are allocated and does not establish priority in the Radio Regulations. (WRC-03)

5.388B In Algeria, Saudi Arabia, Bahrain, Benin, Burkina Faso, Cameroon, Comoros, Côte d'Ivoire, China, Cuba, Djibouti, Egypt, United Arab Emirates, Eritrea, Ethiopia, Gabon, Ghana, India, Iran (Islamic Republic of), Israel, the Libyan Arab Jamahiriya, Jordan, Kenya, Kuwait, Mali, Morocco, Mauritania, Nigeria, Oman, Uganda, Qatar, the Syrian Arab Republic, Senegal, Singapore, Sudan, Tanzania, Chad, Togo, Tunisia, Yemen, Zambia and Zimbabwe, for the purpose of protecting fixed and mobile services, including IMT-2000 mobile stations, in their territories from co-channel interference, a high-altitude platform station (HAPS) operating as an IMT-2000 base station in neighboring countries, in the bands referred to in No. **5.388A**, shall not exceed a co-channel power flux-density of -127 dB (W/(m^2 · MHz)) at the Earth's surface outside a country's borders unless explicit agreement of the affected administration is provided at the time of the notification of HAPS. (WRC-03)

Figure 2.4 Example footnotes to frequency allocation in the ITU Radio Regulations.

IMT-2000

In the case of IMT-2000 the bands that should be used are those identified in the relevant footnotes as shown earlier in this section. In the case of IMT-2000 there are still possibilities to choose different bands, but most commonly the first network deployments were in the

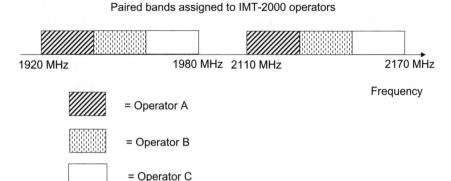

Figure 2.5 Example of how spectrum is made available for mobile networks.

paired bands 1920–1980 MHz and 2110–2170 MHz. National administrations would make spectrum available from those bands for IMT-2000 networks by *assigning* portions of the bands for operators. This is illustrated in Figure 2.5, where we see that the spectrum allocated to Mobile Service and identified for IMT-2000 by the ITU-R is assigned to mobile operators by the national administration.

Dedicated versus shared spectrum

In this example each operator would have *dedicated spectrum* assigned to his network. Dedicated spectrum allows that the radio environment is well known from the interference point of view which in turn allows well-defined Quality of Service (QoS). It is also possible that *shared spectrum* is used. In that case there are one or more spectrum bands that are used by several users of one or more technologies either without any coordination or with some level of coordination. In such usage the radio environment is not precisely known and the interference levels can vary. Operation in such environment may need special measures with respect to protocols, and the QoS is not necessarily guaranteed. The best-known band shared by various technologies is the 2.4 GHz band that is today used by WLANs, and a number of other short range communication technologies.

Global harmonization

In the past it was typical that spectrum usage was different in different countries and the intention was to protect the national industry producing radio equipment. At some point in time the benefits of larger markets were realized and the efforts towards *spectrum harmonization* both regionally and globally were initiated. Regional organizations *designated* specific spectrum bands for specific technologies. The best-known example for this is the GSM spectrum allocation in Europe. Today, on top of this strategy, competition between providers of a service is introduced. This means the allocation of spectrum bands by observing the neutrality of technology, leaving the choice of technology to operators and markets. Of course even in this situation there needs to be some coordination so that the interference levels do not become harmful towards any usage.

Most of the radio spectrum is used for *communication*, either public or private, civil or military, terrestrial or satellite, fixed or mobile, local or wide range, one way or bidirectional. There are many ways to categorize, but typically in modern radio communication there are the current trends for higher bit rates and capacity. As a consequence new bands and wider bandwidths are required. In a situation where there are no unused and available bands, this poses a demanding challenge both for technologies and spectrum management.

IMT-Advanced

One aspect in finding suitable spectrum for a certain purpose is to determine what are the required characteristics of the particular use. It is an issue related to the *propagation characteristics* of electromagnetic waves but it is also about the *required bandwidth*. In some cases the LOS is needed, sometimes there is a need to maximize the distance in a variable terrain. In general the attenuation increases as the frequency is increased. On the other hand, lower bands are very much occupied as they have been taken in use first and they do not offer the possibility for wide bandwidths.

The aspects influencing the spectrum range preference of IMT-Advanced are as follows.

- Target peak data rates and the achievable spectral efficiency: these indicate required bandwidths. Certain bandwidths are feasible only above certain frequencies.

- Mobility: mobile applications require sufficiently low frequencies.

- Coverage range: lower frequencies allow wider range. Wider range allows lower site density.

- Device power consumption: higher frequencies cause higher power consumption.

- Availability and cost of components: higher frequencies mean higher costs of components.

The conclusion was that the new bands should be below 6 GHz for IMT-Advanced; see Report ITU-R M.2074.

Refarming of spectrum

If a new usage is to be accommodated into the radio spectrum, it either means that some of the current usage needs to be moved to another band or the new usage needs to be able to share the band with the original usage. The former case, called *refarming*, usually means that changes are needed to the original equipment and there are costs involved, unless the original usage will come to an end. In the second case the *sharing and compatibility* issues have to be carefully studied and as an outcome there may be technical or operational restrictions to one or both usages. Such sharing and compatibility studies are normally made within the ITU or in national administrations before decisions are taken and allocations are changed.

If there is a need to find and make available new bands, especially in a global manner, and if that requires changes in the RR, the whole process is typically long, taking several years. Therefore it is essential that there are methods and tools that help in estimating the future spectrum requirements. Understanding the current spectrum use is also helpful.

2.3   Radio Communication Services

This section introduces a wide variety of radio communication services that are classified by the ITU-R definition. These services use different frequencies by taking advantage of the characteristics of each frequency.

2.3.1   Mobile service

Mobile service (MS) is a radio communication service between mobile and land stations, or between mobile stations. The land station is defined as a station in the mobile service not intended to be used while in motion. The mobile station is defined as a station in the mobile service intended to be used while in motion or during halts at unspecified points.

Along with the shift to an information-oriented society, there has been an increasing demand for the ability to communicate with anyone, anytime and anywhere: in a car, on a ship, or on a street, free from the constraints of geographical locations of telecommunication facilities. Mobile communication is a means to respond to that demand and is extensively employed. Major forms of mobile communication services include automobile communications, radio pagers, cellular mobile phones, Professional Mobile Radio (PMR), Personal Handy-phone Systems (PHSs), and WLANs. Initially, the UHF band was used because it enabled the transmission of a large volume of information with small-sized antennas and transceivers. In recent years, however, microwaves (SHF) have been increasingly used because of the need to transfer a larger quantity of information in order to support multimedia communications and high-speed Internet access services. Since SHF features more rectilinear and less diffractive propagation in comparison with UHF, it overcomes the drawbacks of high frequencies by means of digital technology and diversity and has facilitated the use of high-frequency bands.

2.3.2   Broadcasting service

Broadcasting service (BS) is a radio communication service in which the transmissions are intended for direct reception by the general public. This service includes sound transmissions, TV transmissions and other types of transmission.

Compared to the wired communication, wireless communication is more suitable for public broadcasting services that transmit signals to an unspecified number of people. Broadcasting media include voice (radio) and video (television).

The AM radio broadcasting uses MF waves as they are stable enough to reach a long distance and can operate with simple receivers, although large transmitters and transmission antennas are still necessary.

For international radio broadcasting, HF waves are used. Their wavelengths of 10 to 100 m allow the signals to be reflected time and again between the ground and the ionosphere formed at the altitude of 200 to 400 km and finally to reach the far side of the earth.

For the FM radio and TV broadcasting services, VHF waves are employed as they can transmit a larger quantity of information compared to microwaves.

2.3.3   Fixed service

Fixed service (FS) is a radio communication service between specified fixed points. Fixed services mainly include digital terrestrial link systems and digital fixed access communication systems which use microwaves for transmission. A digital terrestrial link system is a fixed point-to-point wireless communication system, for which 4, 5 and 6 GHz bands are preferentially used, besides bands at 10 to 20 GHz and even higher. A digital terrestrial link system has a standard link distance of approximately 50 km and realizes a capacity ranging from 150 to 300 Mbit/s per system.

A digital fixed access communication system is a fixed wireless communication system for subscriber lines using microwave bands. This system uses 22, 26 and 38 GHz bands to realize transmission speeds of several to several dozens of Mbit/s. In places with a high user density such as an urban area, point-to-multipoint connection is employed to connect base stations with subscribers. In contrast, in areas with less traffic demands such as a provincial city, point-to-point connection is applied as a more suitable connection method. Recently, IEEE 802.16a, which has become an IEEE standard for wireless metropolitan area network (MAN), has been attracting attention as an affordable means to offer broadband connection services.

2.3.4   Fixed and mobile satellite services

Satellite communication is a wireless communication system in which radio stations (earth stations) communicate with each other via a radio transponder (space station) installed on an artificial satellite. Since a transponder located at a significantly high altitude can cover a huge area, satellite communications feature an extensive service area and a high broadcasting ability in addition to the advantages of terrestrial wireless communications. This is made possible by the signals radiated from the space station which can cover a wide area and can communicate with a large number of base stations scattered in the service area. However, frequencies at 10 GHz and above are subject to constraints such as attenuation due to precipitation. The transmission quality of satellite communications is also limited by the weak received power due to its significantly long transmission path and resultant propagation loss. In addition, there are some drawbacks in satellite communications. First, satellite communications require large equipment such as high-power transmitters, high-gain (large-aperture) antennas and low-noise receivers. Second, delay becomes problematic in geostationary satellite communications. It takes 0.24 s for signals to make a round trip to a satellite in order to travel through the minimum distance of approximately 72 000 km. Third, the number of satellites and the amount of frequency spectrum in a geostationary orbit are limited. Fourth, a satellite has a limited lifetime and it requires high reliability. In order to avoid the drawbacks mentioned above, satellites designed to operate on a lower orbit (low earth orbit satellites) have been put to practical use.

Satellite bands

As shown in Table 2.2, the frequency bands that satellite communications can use have been internationally specified for fixed, mobile and broadcasting satellites. The channel from the earth station to the satellite is called the uplink, while the reverse channel is called the downlink. A pair of frequency bands is used for the uplink and the downlink.

As lower frequencies can output high power more easily and have less attenuation in signal transmission, the lower frequency is used on the downlink in order to reduce the burden on the satellite transmitter.

Table 2.2   Frequency bands for satellite communications.

Name	Frequency band uplink (GHz)	Frequency band downlink (GHz)	Applicable communications
L band	1.6	1.5	Mobile
S band	2.6	2.5	Mobile, broadcasting
C band	6.0	4.0	Fixed, broadcasting
Ku band	14.0	12.0	Fixed, broadcasting
Ka band	30.0	20.0	Fixed, mobile, broadcasting

C-band frequencies have been used since the first days of satellite communications. This is because C-band frequencies in the 1 to 10 GHz band are less susceptible to attenuation and cosmic noise thus having been proven, to be well suited for terrestrial wireless communications. However, there is a shortage of spectrum in the C-band due to an increasing number of C-band systems while more spectrum is necessary to launch new services such as broadcasting. Therefore, higher frequency bands such as the Ku- and Ka-bands are now increasingly utilized for satellite communications as they have more frequencies available and can provide a larger transmission capacity.

Fixed versus mobile

Fixed satellite service (FSS) can be used to provide an intermediate link. This means that satellite communication can be utilized as a link channel in order to enhance the reliability of telecommunication networks. Furthermore, satellite communication is able to accommodate fluctuations in traffic volume flexibly by taking advantage of its adaptable channel settings. In addition, making good use of its wide area coverage and swiftness in the channel setting, satellite communication can cover the sections reaching places such as isolated islands, where terrestrial communication systems can hardly be applied.

Mobile satellite service (MSS) can be used to provide temporary channels and mobile communication channels, again by taking advantage of its swift channel setting and wide area coverage. Vehicle-mounted stations also fall in this application category.

2.4   Radio Communication Systems

In this section, several latest radio communication systems are introduced such as a cellular system, WLAN, terrestrial broadcasting and short-range communications.

2.4.1   Cellular systems

This section introduces the cellular systems, which provide mobile communication services. The cellular concept, which is a key technology for mobile communication, is also introduced.

Network structure of mobile communications services

A network that provides mobile communications services such as mobile telephony consists basically of two parts: the core network and the radio access network. The radio access network directly transmits and receives signals to and from mobile users. On the other hand, the core network is designed to interconnect and communicate with other telephone networks, the Integrated Services Digital Network (ISDN), and the Internet. Figure 2.6 depicts a detailed structure of a mobile communications network.

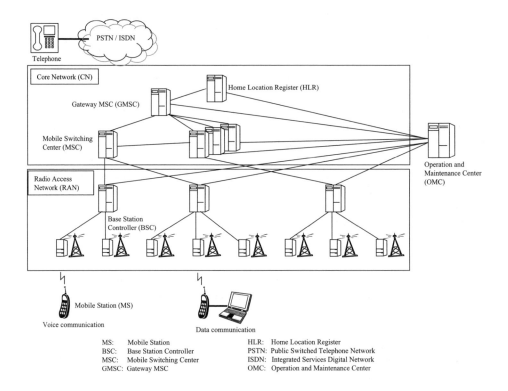

MS:	Mobile Station	HLR:	Home Location Register
BSC:	Base Station Controller	PSTN:	Public Switched Telephone Network
MSC:	Mobile Switching Center	ISDN:	Integrated Services Digital Network
GMSC:	Gateway MSC	OMC:	Operation and Maintenance Center

Figure 2.6 Structure of a mobile communications network.

Base stations

In a mobile communications network, a number of equipment units called base stations (BSs) are installed nationwide at certain intervals to allow mobile phones to connect to the network, no matter where they are. A mobile phone in communication transmits and receives signals to and from the antennas of base stations located in the vicinity.

A base station is composed mainly of antennas and major systems (control systems). The antennas of a base station are usually set up in an open place with a fine view such as the rooftop of a building. The transmitter and receiver (transceiver) systems are normally installed near the antennas on the roof of the building or inside the building. When a base station is installed in a mountain area without any building nearby, it is usually housed in an office exclusively built for the base station.

Cellular system

A base station covers a local area of a certain size to realize transmission and reception of mobile communications signals. This area is called a cell. A base station is usually installed in a way that maximizes the cell coverage reached by its signals in consideration of the surrounding environment such as buildings. The ideal location of base stations is thoroughly investigated so that multiple cells can cover the ground efficiently. The radius of a cell ranges from several hundred meters (e.g. an urban area with a high building and population density) to several dozens of kilometers (e.g. a sparsely populated rural area). In some countries, tens of thousands of base stations are necessary to cover the whole land. A mobile communications system is also called a cellular system.

In providing mobile communications services, efficient use of spectrum is very important in order to offer services to as many subscribers as possible, using such limited spectrum. For this reason, the ability to use the same spectrum repeatedly in geographically separated places is perceived as being mandatory.

Ad hoc concept

Mobile communications are divided into two types: communications between mobile terminals without and with the mediation of base stations. The former type is typical for ad hoc networks and needs a distributed access control protocol to the radio medium to ensure that the transmitted data will not suffer from interference from other users during communication. Ad hoc networks are the secondary choice, since they are not efficient in spectrum use. Systems operating under mediation of a base station applying central control are known to be more efficient since the same spectrum can be allocated to different base stations (distant enough) without causing harmful interference to each other.

Single and multizone concepts

In mobile communication systems that involve base stations, a service area is covered in two ways: one is to use only one base station and the other is to use two or more base stations such as in a cellular mobile radio system as shown in Figure 2.7(a) and (b), respectively.

When an entire area is exclusively covered by radio signals from a single base station, the method of service area coverage is called the single zone configuration. If the area is covered by multiple base stations, we have a multizone configuration. A system with a single zone configuration is simple, since it comprises only mobile terminals, the transmitter/receiver facilities in the base station and the communication link between the base station and the switch connecting to a wired network. Such a system requires high transmission power for the base station and mobile terminals in order to cover its wide service area. A sufficiently long distance, therefore, must be secured to permit spectrum reuse between two single zone systems.

To be able to carry the communications traffic generated per area unit in a large single zone, much more frequency spectrum is required than needed with multizone configurations. Since the amount of spectrum licensed to a mobile network operator is always limited in

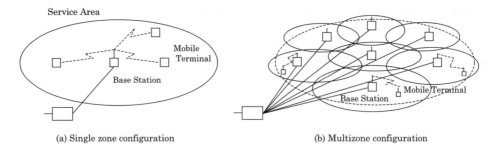

(a) Single zone configuration   (b) Multizone configuration

Figure 2.7 Single and multizone configurations of a cellular system.

bandwidth, all the public mobile radio networks operated nowadays apply the multizone configuration. Then the traffic capacity (bit/s per area unit) under a limited bandwidth condition can be adjusted to the need perfectly.

Cellular concept

Multizone configurations have the following characteristics, making them preferable to single zone configurations:

- improved spectrum efficiency, since the same spectrum can be used for different communications performed in different zones that are sufficiently spaced from each other to avoid mutual interference. The spectrum efficiency increases with decreased size of the zone;

- ability to develop a service area flexibly in size and shape so that it meets the requirements while maintaining a good QoS level by configuring the area as a combination of multiple small zones;

- consumption of less transmission power;

- requirement of more information exchange between base and mobile stations for status monitoring and connection control in order to maintain paging efficiency and voice call continuity, resulting in a complex system structure.

In spite of its drawback of complicated system structure, the multizone configuration has been utilized in mobile radio systems today as it can cover a wide service area and realize a large capacity system with higher spectrum efficiency. In the multizone configuration, a service area is composed of multiple zones, each of which centers around a base station that resembles a cell which makes up living things with a nucleus at its center. This is why the zone is called the cell, and the multizone-based mobile communication system is called the cellular system.

In a cellular system, the number and the size of the cells are determined by parameters such as type of terrain, radio propagation characteristics, traffic volume distribution and economical efficiency. Especially, the placement of base stations in order to cover the service area thoroughly, and the identification of base stations that share the same spectrum must be carefully decided to maximize the spectrum efficiency.

Macro, micro and pico cells

As the area covered by each base station becomes wider, the spectrum efficiency deteriorates although the number of base stations can be decreased. On the other hand, as the coverage area gets smaller, an increased number of users can be served within a limited spectrum but an increased number of base stations is needed. Cells that cover those service areas are categorized as macro cells, micro cells or pico cells in descending order of their size. Macro cells cover areas of several kilometers to several hundred meters in diameter. On the other hand, micro cells, which cover areas with a diameter of several dozen to several hundred meters, require a well-planned installation of base stations to cover those small areas. In addition, pico cells to cover very small areas (like inside an elevator) are introduced in order to provide efficient coverage for indoor areas and heavily populated areas.

Figure 2.8 gives an example of a linear (single-layer) cell configuration, where the same frequency channels are reused every four cells. The mutual interference of co-channel cells in this configuration is low enough to avoid harmful interference, since the path loss for signals propagation across the two intermediate cells is sufficient to bring the co-channel signal energy below a desired margin. When a service area is widely spread, in contrast, as in a mobile radio system, a linear configuration is not sufficient to provide the desired radio coverage. The cells are arranged, therefore, over the plane surface in three dimensions, introducing cell layers with macro, micro and pico cells. A cell in a mobile system usually has a somewhat random shape as radio propagation depends on terrain and topographical features. Modeling of a mobile system usually assumes a service area with a uniform terrain and topographical features where base stations are placed at regular intervals.

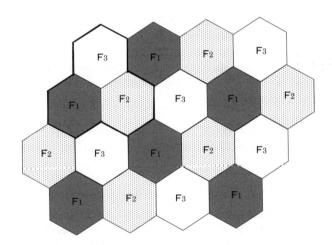

Figure 2.8 Example of linear cell configuration.

Omnidirectional cells and sectored cells

A hexagon is said to be the most suitable shape to represent a zone. In reality, however, it is impossible to achieve hexagonal cells owing to constraints of base station positioning and non-uniform radio propagation owing to inhomogeneous terrain. Therefore, cells are being planned mutually overlapping on their edges so that mobile terminals can be served continuously even when moving across cell borders, and the location to switch service from one cell to the next (called handover) becomes a small area, instead of a line bordering the cells. When an omnidirectional antenna is installed at the center of a cell to cover 360 degrees (all directions), that cell is called an omnidirectional cell. An omnidirectional cell is used mainly in a rural area or in an isolated area with limited traffic. In contrast, in an area with much traffic, sectorized cells are used such that each hexagonal cell is divided radially from the center into three or six sectors each served by an antenna with sector directivity. A sectorized cell is able to serve three or even six sectors without requiring a respective number of base station locations. Instead, only one location is required at the center of the cell, where all the base stations, each serving one sector, are grouped. Sectorization of cells reduces co-channel interference in general and, if well designed, increases spectrum efficiency.

2.4.2   Wireless local area networks

The personal computer (PC) market has witnessed a continuous shift from business use to personal use as PCs have rapidly become widespread in households as well as in offices. At homes and offices today, where a number of computers and peripheral devices exist, local area networks (LANs) have been introduced to connect those devices efficiently. Wireless LANs (WLANs) have gained popularity, since they permit simple and flexible networking between computer and peripheral devices.

WLAN systems have the following characteristics.

- WLANs can provide flexibility and scalability in building LANs in existing offices and houses, eliminating the cumbersome cabling.

- Wireless communication media can improve the convenience of network access, which is enabled from anywhere in offices and households without compromising terminal portability and user mobility.

- Technology evolution allows for faster WLANs, able to support multimedia contents transmission such as large-volume video data.

IEEE WG 802.11

The leading organization in developing standards for WLANs is the Institute of Electrical and Electronics Engineers (IEEE). Working Group IEEE 802.11 has been developing standards for WLANs since 1990. As shown in Table 2.3, IEEE 802.11 standard was released in 1997 based on a common medium access control (MAC) protocol with three options for the physical medium signal transmission:

(1) direct sequence spread spectrum (DSSS)

(2) frequency hopping spread spectrum (FHSS)

(3) infrared (IR).

Table 2.3   Major standards for WLANs by IEEE 802.11 working groups.

Standard	Frequency	Max data rate	Completion year
802.11	2.4 GHz, Infrared	1 Mbit/s, 2 Mbit/s	1997
802.11a	5 GHz	54 Mbit/s	1999
802.11b	2.4 GHz	1 Mbit/s	1999
802.11g	2.4 GHz	54 Mbit/s	2003
802.11n	5 GHz, 2.4 GHz	600 Mbit/s	2007

Options (1) and (2) operate in the Industrial, Scientific and Medical (ISM) frequency band at 2.4 GHz and feature a transmission distance of approximately 100 m at a transmission speed of 1 or 2 Mbit/s.

It took seven years to agree on the IEEE 802.11 standard. After that, to boost the introduction of wireless broadband multi-media for Internet access, standards IEEE 802.11a and 802.11b were standardized in the same year (1999) as shown in Table 2.3, followed by the IEEE 802.11g in 2003. Standard IEEE 802.11a operates in the 5 GHz band, globally allocated by ITU-R to high-speed wireless access. Standards 802.11a and 802.11g, both, apply Orthogonal Frequency Division Multiplexing (OFDM) and specify a set of modulation and coding schemes permitting link adaptation to achieve data rates of between 6 and 54 Mbit/s according to the current radio conditions. IEEE 802.11g operates at 2.4 GHz. See Figure 2.9 for the current IEEE 802.11 standards.

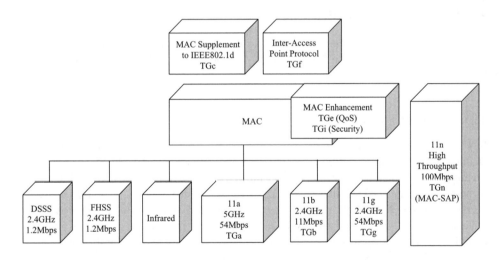

Figure 2.9 Structure of wireless LAN (IEEE 802.11) standards.

In September 2003 Task Group (TG) IEEE 802.11n started to develop a system with 'a quarter Gbit/s' user data rate in mind. The first (pre-standard) chips are already available on the market. WLANs have been enablers for wireless broadband communication for nomadic applications. Some people see the standard IEEE 802.16 (WiMAX), see Section 1.1.2, as a second generation WLAN that also aims to cover mobile broadband services.

Table 2.4   Terrestrial digital broadcasting schemes in Japan, Europe and North America[a].

	ISDB-T	DVB-T	ATSC
Country/region	Japan	Europe	North America
Organization	ARIB	ETSI	ATSC
Video coding	MPEG-2	MPEG-2	MPEG-2
Voice coding	MPEG-2 AAC	MPEG-2 Audio Layer II	Dolby AC-3
Multiplexing	MPEG-2 system	MPEG-2 system	MPEG-2 system
Transmission	OFDM multi-carrier	COFDM multi-carrier	Single carrier
Modulation	DQPSK, QPSK 16QAM, 64QAM	QPSK 16QAM, 64QAM	8-VSB
Bit rate	23 Mbit/s	31 Mbit/s	19 Mbit/s
Bandwidth	6 MHz	6/7/8 MHz	6 MHz
Segmentation	Yes	No	No
Broadcasting	Yes (one segment)	Yes (one channel)	No
Countries/regions adopting the scheme	Japan, Brazil	European countries, Australia, India, South Africa	North America, Korea, Mexico

[a]For acronyms, see Appendix C.1.

2.4.3   Terrestrial broadcasting

This section introduces the current feature and technology for terrestrial broadcasting. First, terrestrial digital broadcasting that is currently replacing analog broadcasting is introduced. Second, a few examples of new mobile broadcasting technologies are introduced.

Terrestrial digital broadcasting

In order to realize efficient use of spectrum, the terrestrial broadcasting has evolved from analog (terrestrial analog TV broadcasting) to digital. While terrestrial Digital Video Broadcasting (DVB-T) has been developed and introduced in Europe and Advanced Television Systems Committee (ATSC) in North America, the Integrated Services Digital Broadcasting-Terrestrial (ISDB-T) has been standardized and introduced for terrestrial digital broadcasting in Japan. Two of these standards are based on OFDM transmission, offering a number of modulation and coding schemes to be able to adapt the data rate to the different services offered, such as TV broadcasting, mobile service, Internet access and so on.

Unlike satellite broadcasting, terrestrial broadcasting is sensitive to multipath (or delayed signals) fading caused by the radio signals reflected from terrain objects such as building surfaces. Multipath fading appears as a ghost in analog broadcasting. OFDM is resistant to multipath fading and this allows OFDM to handle non-dominant signals as multipath signals and to use the single frequency network (SFN) scheme in order to transmit signals with the same frequency in adjacent areas. Thus, OFDM can use the spectrum efficiently.

The main characteristics of the three digital broadcasting standards mentioned are shown in Table 2.4. We note that DVB-T is deployed in many countries throughout the world.

Multicast communication

Multicast is a technology to transmit the same data to multiple parties in a network. A mobile phone makes a call by specifying a called party and establishing a channel to the party. Such a communication method, where data is exchanged with a single party at one time, is called unicast communication. In contrast, multicast communication enables simultaneous transmission of the same data to and from multiple parties. Multicast has attracted attention for its use in Internet, and it has been a target of research for a long time.

This trend has been driven by the need for the application of terminals that can support richer media as the data rates of digital channels available to terminals have increased year by year. However, the bandwidths of those channels do not increase in proportion to the increase in the date speeds achieved on the channels. For example, mobile terminals today have a video telephony function to send images besides voice conversation, which is less data intensive. Some services available today permit the video telephony function to send the same contents to multiple receive terminals.

When transmitting a multi-destination message in unicast communication mode, a point-to-point connection is established from the sender of a video to each terminal that is to receive the video. Therefore, if g terminals are to receive the video, the total bandwidth required for sending the multi-destination video message in the unicast mode is g times the bandwidth needed for a single video delivery. This is a very inefficient use of spectrum.

Mechanisms are known to efficiently realize acknowledged multicast transmission. In a wired network, a sender transmits so-called multicast packets specifying multiple destinations to a router that in turn transfers the packets to the multiple destinations specified (or to another router), or copies the packets and sends them to different routers or networks. A multicast tree is being established for that purpose permitting any receiver to acknowledge the data received to the original sender.

Broadcast is another addressing mechanism that enables all terminals in a network to receive the same data although the information flow is one-way communication from one sender to multiple receivers. With broadcast, receivers do not respond to the sender, e.g. to acknowledge the data received, thus, the sender gives no thought to the situation of the receivers.

Mobile multicast

Acknowledged multicast is difficult to implement in mobile systems.

A terminal may move when it is receiving data and may travel far enough to move into an area served by another base station, causing a change in the network configuration. When that happens, a handover should be conducted to identify the base station which is now serving the terminal instead of the previous one and to determine the route to transmit data to the terminal. If the handover is not performed properly, the terminal may drop the data sent to it.

Several multicast protocols have been developed for mobile terminals. Besides network coding, a protocol called Multimedia Broadcast/Multicast Service (MBMS) is worth mentioning that has been developed by 3GPP. Instead of unicast, MBMS applies point-to-multipoint connections, where a separate connection is established for each session between the sender and a receiver involved. MBMS, thereby, increases the number of channels that can be provided in a network. This is because MBMS allows multicast to be performed to multiple users who request the same contents by adapting the combination of channels dynamically in accordance with the request from each cell. MBMS is part of the evolved W-CDMA standard of 3GPP. MBMS will be available for both GSM/EDGE and W-CDMA. 3GPP2 has standardized almost the same technology called BroadCast and MultiCast Services (BCMCS).

2.4.4   Short-range communications

A major short-range wireless communication system is the Bluetooth explained here.

Bluetooth was originally developed as a technology to 'interconnect mobile terminals and peripheral devices that exist within a range of only several meters without cables'. Bluetooth operates in the 2.4 GHz ISM band that is globally harmonized in European countries, in the United States, and in Japan.

Around 1996, before the Internet e-mail connection via mobile phones became popular, the mobile data communication was performed by connecting a PDA or a dedicated terminal to the mobile phone. In 1997 the packet communication was started with mobile phones. It was around this time when the mobile data communication market was established. Nowadays, a variety of mobile Internet applications are available that enables services combining the mobile phone with the camera. In addition, many external devices are connected to the mobile phone. Smooth utilization of such mobile connection environments has required a short-range wireless access technology such as Bluetooth.

Bluetooth was first introduced by the Bluetooth Special Interest Group (SIG) in 1998. After that the scope of Bluetooth was expanded to embrace any kind of mobile terminals. This is why Bluetooth was specified as an inter-industrial technical standard across the PC and mobile phone industries. In 2007, the SIG membership has increased from 15 companies at the time of its foundation to more than 9000 companies and organizations.

3

Spectrum Requirement Calculation for IMT-2000

Hideaki Takagi

IMT-2000 defines the third generation (3G) mobile systems which started service around the early 2000s. Two notable examples of air interface are:

- Wideband Code Division Multiple Access (W-CDMA)

- cdma2000.

European standards of IMT-2000 are called Universal Mobile Telecommunications System (UMTS), while Freedom of Mobile Multimedia Access (FOMA) is the brand name for the 3G services launched in 2001 by NTT DoCoMo, Japan. In the United States and some Asian countries, cdma2000 was deployed in 2002 that is much more flexible in spectrum deployment, owing to its 'narrow' channel bandwidth of 1.25 MHz compared to the 5 MHz bandwidth of UMTS.

The methodology for terrestrial spectrum requirement calculation based on a blended second generation (2G) and IMT-2000 technology networks is described in Recommendation ITU-R M.1390. Using that methodology, the spectrum required to carry the traffic projected for the year 2010 is estimated in Report ITU-R M.2023 using an MS Excel implementation of the calculation methodology. These documents were prepared for the World Radiocommunication Conference 2000 (WRC-2000).

In this chapter, we present the methodology for the spectrum requirement calculation for IMT-2000 by following Recommendation ITU-R M.1390 along with the market forecast numerical data and other input parameter values given in Report ITU-R M.2023. Section 3.1 describes the calculation model consisting of environments, services and direction of links. The flow chart of methodology is also shown. Section 3.2 introduces input parameters regarding geography, personal traffic demand, radio systems, required service quality and

Spectrum Requirement Planning in Wireless Communications   Edited by Hideaki Takagi and Bernhard H. Walke
© 2008 John Wiley & Sons, Ltd

final adjustment. The methodology for calculating the offered traffic, determining the required spectrum and making final adjustment is shown in Section 3.3, in which application of Erlang-B and Erlang-C formulas is explained in detail. Finally, Section 3.4 remarks on the spectrum results determined at the WRC-2000. It also refers to examples of application of the same methodology to other real-world systems.

We note that the model and method in the methodology for IMT-2000, presented in this chapter, are quite different from those in the methodology for IMT-Advanced to be presented in Chapter 4 and onward. Therefore, we have not tried to unify the terminology and notation of symbols between this chapter and the subsequent chapters.

3.1   Model

As a model for calculating the required spectrum for IMT-2000, we assume that the traffic demands are generated by users of different services present in each cell. The characterization of a cell by its area, the density of user population, and their traffic demand is called the environment. The traffic generated in each cell is transmitted over the radio channel in uplink or downlink direction with service-dependent bit rate. Therefore, the input, intermediate and final parameters used in the calculation are denoted with indices for environment, service and link direction. In addition, these parameter values may be different in different regions of the world, but the index for region is implicit.

3.1.1   Environments

The *environment* characterizes the cell with respect to its size, offered traffic and radio system parameters. The environments are defined by a combination of the density attributes and the mobility attributes:

$$\{\text{environment}\} = \{\text{density}\} \times \{\text{mobility}\}. \tag{3.1}$$

The *density* attributes are categorized as follows:

- high-density

- urban

- suburban

- rural.

The *mobility* attributes are related to the speed of the movement of users. They are categorized as follows:

- in-building: stationary

- pedestrian: 3–10 km/h

- vehicular: about 50 km/h.

The environment is denoted by index e in the parameters.

Hence, there are $4 \times 3 = 12$ different environments. For the spectrum requirement estimate in Report ITU-R M.2023, the following three environments are considered.

- High-density in-building

  This environment represents a *central business district* (CBD). Since most users may be sitting and making many phone calls in the office, the cell area will be rather small. For example, the cell in CBD is taken to have an area of 5000 m^2. If the cell is circular, the radius is about 40 m. If it is a square, the size is about 70 m $\times$ 70 m.

- Urban pedestrian

  This is representative of highly populated pedestrian areas such as those in roadways and shopping districts where people walk around. Pedestrians may move rather widely, but they do not make as many calls as office workers. For example, the urban pedestrian cell is taken to have an area of 312 000 m^2. This is considered to be a circle with radius of 315 m, or a square of 560 m $\times$ 560 m.

- Urban vehicular

  Urban vehicular cells are large while calls occur much less often than with people. If a hexagon with radius of 1000 m is divided into three sectors, the cell area is 866 000 m^2.

3.1.2 Services

Recommendation ITU-R M.1390 considers *bearer services* which are defined in terms of the type of transmission (circuit-switched or packet-switched) and the transmitted user bit rate. They are not *application services* defined in terms of the use of bit transmission such as voice, e-mail, video conferencing, etc.[1]

The following services are considered:

- speech (S): toll quality voice service

- simple message (SM)

- switched data (SD)

- medium multimedia (MMM)

- high multimedia (HMM)

- high interactive multimedia (HIMM).

The service is denoted by index s in the parameters.

We note that S, SM and SD are first generation (1G) and 2G services, while MMM, HMM and HIMM are considered as new IMT-2000 services.

MMM and HMM are asymmetric multimedia services such that more traffic flows in one direction than in the other direction. In Report ITU-R M.2023, the higher rate of flow is on the downlink, while the lower rate of flow is on the uplink. Some examples of such asymmetric services include file download, Internet browsing, full motion video and non-interactive telemedicine. On the other hand, HIMM is supposed to provide symmetrical multimedia service such that an equal amount of traffic flows in both directions. Examples of such symmetric services are high fidelity audio, videoconferencing and interactive telemedicine.

---

[1]Application services of IMT-2000 are defined in Section 7 of Recommendation ITU-R M.816.

Table 3.1   Characterization of bearer services with net user bit rate in kbit/s.

Service	Type	Uplink bit rate	Downlink bit rate
S	circuit-switched	16	16
SM	packet-switched	14	14
SD	circuit-switched	64	64
MMM	packet-switched	64	384
HMM	packet-switched	128	2000
HIMM	circuit-switched	128	128

Circuit-switched versus packet-switched services

In communication networks, a connection path called a *circuit* is needed when data are sent from the transmitter to the receiver. Usually a circuit consists of the transmitter node (source), several intermediate relaying nodes, the receiver node (destination) and the communication links connecting those nodes. Physically communication links may be made of copper, optical fiber cable or spectrum of electromagnetic wave. The circuit may either be fixed during the communication period once it is established, or it may be created every time data are actually transmitted.

For *circuit-switched* services, before the session of a service can start between two parties, a circuit is established between them. The network resource on this circuit is reserved only for this session. All data exchange of this session takes place over this circuit, while other sessions cannot use the resource on this circuit even when it is not used by this session momentarily. The circuit is closed at the end of the session, when the resource is released. An example of circuit-switched service is the classical telephone system.

For *packet-switched* services, no circuit is set up prior to actually sending data during the session of a service between two parties. A circuit is established only when a block of data, called packets, is transmitted. Therefore, different connection paths may be used during the session. The session is maintained only logically by the pair of source and destination nodes. An example of packet-switched service is the electronic mail in a computer network.

Net user bit rate

The service bit rate of a bearer service shows how many bits are transmitted per unit time (usually a second) for this service. This rate is called the net user bit rate and denoted by $Net\_User\_Bit\_Rate_s$ for service s. It is measured by the unit of kilo ($= 1000$) bits per second (kbit/s).

The type and the net user bit rate for each service assumed in Report ITU-R M.2023 are listed in Table 3.1.

3.1.3   Direction of links

The uplink (UL) is the link from the mobile station to the base station. The downlink (DL) is the link from the base station to the mobile station. The spectrum requirements are calculated separately for uplink and downlink directions owing to the possible asymmetry in the traffic of some services.

The link is denoted by index l in the parameters.

3.1.4 Region

As mentioned in Section 2.2, ITU-R divides the world into three Regions (see Figure 2.2):

- Region 1: Europe, Middle-East and Africa

- Region 2: North and South America

- Region 3: Asia-Pacific.

The input parameters are given and calculations are made separately for different Regions. Therefore, the index of Region is implicit in the parameters below.

3.1.5 Flow chart of methodology for IMT-2000

Figure 3.1 shows the flow chart of spectrum requirement calculation for IMT-2000 presented in this chapter. Such a flow chart is not provided in any ITU documents. This flow chart is newly created for the present book by the author.

In this flow chart, the oval boxes show the input parameters, and the rectangular boxes show the parameters calculated in the methodology.

3.2 Input Parameters

The input parameters used in the spectrum requirement calculation methodology in Recommendation ITU-R M.1390 are here classified into the following categories:

- geographic parameters

    - $\texttt{Cell\_Area}_e$ (km^2/cell)
    - $\texttt{Population\_Density}_e$ (users/km^2)
    - $\texttt{Penetration\_Rate}_{e,s}$ (%)

- personal traffic parameters

    - $\texttt{Busy\_Hour\_Call\_Attempts}_{e,s}$ (calls/h/user)
    - $\texttt{Call\_Duration}_{e,s}$ (s)
    - $\texttt{Activity\_Factor}_{e,s,l}$ (dimensionless)

- radio system parameters

    - $\texttt{Group\_Size}_{e,s}$ (cells/group)
    - $\texttt{Service\_Channel\_Bit\_Rate}_{e,s}$ (kbit/s)
    - $\texttt{Net\_System\_Capability}_{e,s}$ (bit/s/Hz/cell)

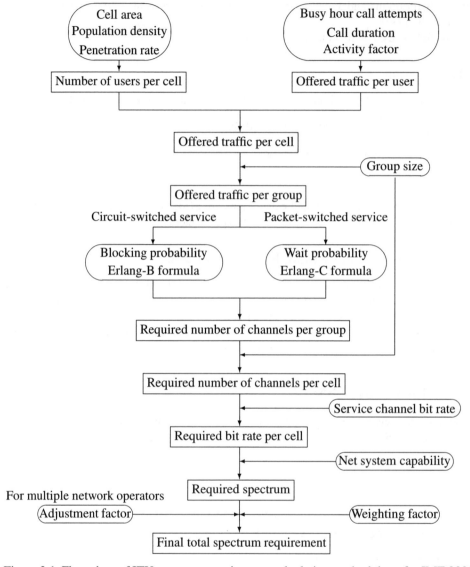

Figure 3.1 Flow chart of ITU spectrum requirement calculation methodology for IMT-2000.

- Quality of Service (QoS) parameters

    - Blocking probability for circuit-switched service (dimensionless)

    - Tail probability of the waiting time for packet-switched service (dimensionless)

- adjustment parameters

    - Weighting factor $\alpha_{e,s}$ (dimensionless)

    - Adjustment factor β (dimensionless)

where the subscripts e, s and l denote the dependence on environments, services and links, respectively.

3.2.1 Geographic parameters

In order to estimate the traffic volume per cell, we first have to know how many potential users reside on average in a cell. The basic data for calculating the average number of potential users present in a cell are the following geographic parameters:

- `Cell_Area`$_e$ (km^2/cell)

- `Population_Density`$_e$ (users/km^2)

- `Penetration_Rate`$_{e,s}$ (%).

In the market forecast data given in Report ITU-R M.2023, the values of geographic parameters depend on the region. From these parameters, we calculate the intermediate parameter `Users_per_Cell`$_{e,s}$ as shown in Section 3.3.1.

Cell area

The input parameter `Cell_Area`$_e$ (km^2/cell) denotes the area of each cell in environment e. Typical shapes of a cell are a circle and a hexagon as shown in Figure 3.2.

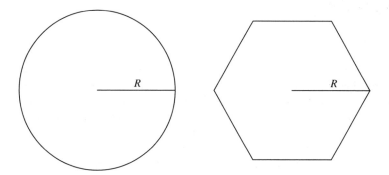

Figure 3.2 Circular and hexagonal cells.

For a circular cell, the area of the cell is given by

$$\texttt{Cell\_Area}_e = \pi R^2 \approx 3.1416 R^2, \tag{3.2}$$

where R (km) is the radius of the circle.

For a hexagonal cell, the area of the cell is given by

$$\texttt{Cell\_Area}_e = \frac{3\sqrt{3}}{2} R^2 \approx 2.5981 R^2, \tag{3.3}$$

where R (km) is the radius (distance from the center to a vertex) of the hexagon.

For a cell which is a sector of a circle/hexagon, the Cell_Area$_e$ must be the area of a full circle/hexagon divided by the number of sectors.

Numerical examples of the cell area are shown by regions in Table 3.2.

Table 3.2   Cell area (km^2/cell).

		Environments	
Cell	CBD	Urban pedestrian	Urban vehicular
Region 1, 2			
Geometry	Circle	Hexagon	Hexagon
R (km)	0.04	0.60	1.00
Sectors	1	3	3
Cell area	0.00503	0.312	0.866
Region 3			
Geometry	Circle	Hexagon	Hexagon
R (km)	0.04	0.60	0.60
Sectors	1	3	3
Cell area	0.00503	0.312	0.312

Population density

The input parameter Population_Density$_e$ (users/km^2) denotes the average number of people/vehicles per unit area in the environment e under consideration. This includes not only the number of inhabitants in the area but also mainly commuters and shoppers, i.e. people and vehicles in the environment at a given point in time for the need of their business or for their leisure.

Numerical examples of the population density in each region are shown in Table 3.3. If we assume a two-floor building, the density of 140 000 people/km^2 corresponds to 70 000 people/km^2/floor, which means on average one person in 14 m^2 of floor space of an office environment. In the urban pedestrian environment, the density is 100 000 people/km^2 for Regions 1 and 3, while it is 75 000 people/km^2 for Region 2. In the urban vehicular environment, the density is 3000 vehicles/km^2.

Table 3.3   Population density (users/km^2).

		Environments	
Region	CBD	Urban pedestrian	Urban vehicular
1	140 000	100 000	3000
2	140 000	75 000	3000
3	140 000	100 000	3000

It is clear that the product

$$\text{Cell\_Area}_e \times \text{Population\_Density}_e \quad \text{(users/cell)}$$

gives the average number of people or vehicles per cell in the environment e.

Penetration rate

Not all the people/vehicles subscribe to service s in environment e. The input parameter $\text{Penetration\_Rate}_{e,s}$ (%, dimensionless) is the ratio of the number of people/vehicles subscribing to service s over the total population in environment e.

It should be noted that the subscription to services is not exclusive. Since users may subscribe to more than one service, it is possible for the sum of the penetration rates over all services in an environment to exceed 100%.

Numerical examples of the penetration rate (common to all regions) are shown in Table 3.4.

Table 3.4 Penetration rate.

Service	Environments		
	CBD (%)	Urban pedestrian (%)	Urban vehicular (%)
S	73	73	73
SM	40	40	40
SD	13	13	13
MMM	15	15	15
HMM	15	15	15
HIMM	25	25	25

3.2.2 Personal traffic parameters

The second component needed to estimate the traffic volume per cell is the amount of traffic generated by an average user in a unit time. Each user's traffic demand is specified by the following parameters:

- $\text{Busy\_Hour\_Call\_Attempts}_{e,s}$ (calls/h/user)

- $\text{Call\_Duration}_{e,s}$ (s)

- $\text{Activity\_Factor}_{e,s,l}$ (dimensionless).

It is assumed in Report ITU-R M.2023 that the values of these personal traffic parameters are common to all regions. From these parameters, we calculate the intermediate parameter $\text{Offered\_Traffic\_per\_User}_{e,s,l}$ in Section 3.3.1.

Busy Hour Call Attempts (BHCA)

The input parameter Busy_Hour_Call_Attempts$_{e,s}$ (calls/h/user) is defined as the average number of calls attempted by an arbitrary user of service s per hour during the busy hour in environment e. For packet-switched services, a call is understood as a session.

Numerical examples of the BHCAs (common to all regions) are shown in Table 3.5.

Table 3.5   Busy hour call attempts (calls/h/user).

Service	CBD	Urban pedestrian	Urban vehicular
		Environments	
S	3	0.8	0.4
SM	0.6	0.3	0.2
SD	0.2	0.2	0.02
MMM	0.5	0.4	0.008
HMM	0.15	0.06	0.008
HIMM	0.14	0.07	0.011

Call duration

The input parameter Call_Duration$_{e,s}$ (s) is defined as the average duration of a call (for circuit-switched service) or of a session (for packet-switched service) during the busy hour.

Numerical examples of the call duration (common to all regions) are shown in Table 3.6.

Table 3.6   Call duration (s).

Service	CBD	Urban pedestrian	Urban vehicular
		Environments	
S	180	120	120
SM	3	3	3
SD	156	156	156
MMM	3000	3000	3000
HMM	3000	3000	3000
HIMM	120	120	120

Activity factor

The input parameter Activity_Factor$_{e,s,l}$ (dimensionless) is defined as the percentage of time during which the link in direction l is actually used during the call of service s in environment e. For example, if voice is transmitted only when the user speaks, or if the packet transmission is bursty, the transmission is only active during a relatively small portion of time.

Numerical examples of the activity factor are shown in Table 3.7.

Table 3.7 Activity factor (dimensionless).

Service	Link	Environments		
		CBD	Urban pedestrian	Urban vehicular
S		0.5	0.5	0.5
SM		1	1	1
SD		1	1	1
MMM	UL	0.00285	0.00285	0.00285
	DL	0.015	0.015	0.015
HMM	UL	0.00285	0.00285	0.00285
	DL	0.015	0.015	0.015
HIMM		1	1	1

3.2.3 Radio system parameters

The following parameters related to the radio system are also used as input to the calculation:

- $\texttt{Group\_Size}_{e,s}$ (cells/group)

- $\texttt{Service\_Channel\_Bit\_Rate}_{e,s}$ (kbit/s)

- $\texttt{Net\_System\_Capability}_{e,s}$ (bit/s/Hz/cell).

Group size

The input parameter $\texttt{Group\_Size}_{e,s}$ (cells/group) is the number of cells per group. Seven (7) is uniformly chosen for the group size, because it represents the cell in question plus the surrounding six first tier cells in the hexagonal tessellation of a cellular system:

$$\texttt{Group\_Size}_{e,s} = 7. \qquad (3.4)$$

See Figure 2.8 for the illustration.

Service channel bit rate

The *service channel bit rate*, denoted by $\texttt{Service\_Channel\_Bit\_Rate}_{e,s}$ (kbit/s), is the net user bit rate of bearer service s (defined in Section 3.1.2) plus any channel overhead bits (e.g. for signaling and timing) needed to carry that service in environment e. For example, an 80 kbit/s service channel bit rate may be necessary to carry a 64 kbit/s net user bit rate of the SD service. However, it is assumed in Recommendation ITU-R M.1390 that the service channel bit rate is simply the same as the net user bit rate by neglecting the overhead bits. Thus we have

$$\texttt{Service\_Channel\_Bit\_Rate}_{e,s} = \texttt{Net\_User\_Bit\_Rate}_s. \qquad (3.5)$$

Accordingly, the numerical values of the service channel bit rate are copied from Table 3.1 as shown in Table 3.8 for Region 1.

Table 3.8   Service channel bit rate (kbit/s) in Region 1.

Service	Link	CBD	Urban pedestrian	Urban vehicular
			Environments	
S		16	16	16
SM		14	14	14
SD		64	64	64
MMM	UL	64	64	64
	DL	384	384	384
HMM	UL	128	128	128
	DL	2000	2000	2000
HIMM		128	128	128

Net system capability

The parameter Net_System_Capability$_{e,s}$ (bit/s/Hz/cell) denotes the spectrum bandwidth in Herz (Hz) needed to carry the data rate of 1 bit/s per cell. This is used to convert the required bit rate per cell to the required spectrum in Section 3.3.3.

The values of the net system capability may be estimated by means of simulation and validated in deployed systems. For the numerical examples, however, we simply use the values given in Table 3.9.

Table 3.9   Net system capability (bit/s/Hz/cell).

Service	Region 1	Region 2	Region 3
S	0.070	0.100	0.070
SM	0.125	0.150	0.125
SD	0.125	0.150	0.125
MMM	0.125	0.150	0.125
HMM	0.125	0.150	0.125
HIMM	0.125	0.150	0.125

3.3 Methodology

The methodology for determining the spectrum bandwidth that satisfies the QoS requirement by users consists of the following three steps:

(1) calculation of offered traffic

(2) determination of required spectrum

(3) weighting and adjustment.

In the second step, we apply Erlang-B formula to circuit-switched services and Erlang-C formula to packet-switched services.

3.3.1 Calculation of offered traffic

We can calculate the average number of users per cell from the geographic input parameters. In parallel, we can calculate the offered traffic per user from the personal traffic parameters. Multiplying the average number of users per cell with the offered traffic per user yields the offered traffic per cell. Multiplying the offered traffic per cell by the group size (number of cells per group) gives the offered traffic per group.

The following intermediate parameters are calculated:

- `Users_per_Cell`$_{e,s}$ (users/cell)

- `Offered_Traffic_per_User`$_{e,s,l}$ [(calls · seconds)/(user · hour)]

- `Offered_Traffic_per_Cell`$_{e,s,l}$ [(calls · seconds)/(cell · hour)]

- `Offered_Traffic_per_Group`$_{e,s,l}$ (Erlang/group).

Number of users per cell

The parameter `Users_per_Cell`$_{e,s}$ (users/cell) denotes the average number of users who subscribe to service s per cell of environment e. It is given by

$$\text{Users\_per\_Cell}_{e,s} = \text{Cell\_Area}_e \times \text{Population\_Density}_e$$
$$\times \frac{\text{Penetration\_Rate}_{e,s}}{100}. \tag{3.6}$$

The calculated numerical values of the average number of users per cell in each region are shown in Table 3.10.

Offered traffic per user

The parameter `Offered_Traffic_per_User`$_{e,s,l}$ [(calls · seconds)/(user · hour)] gives the traffic volume offered by a user of service s in environment e on link l. It is calculated by

$$\text{Offered\_Traffic\_per\_User}_{e,s,l} = \text{Busy\_Hour\_Call\_Attempts}_{e,s}$$
$$\times \text{Call\_Dduration}_{e,s}$$
$$\times \text{Activity\_Factor}_{e,s,l}. \tag{3.7}$$

The value of `Offered_Traffic_per_User`$_{e,s,l}$ represents how many seconds per hour is used by the traffic from each user of service s in environment e on link l.

The calculated numerical values of the offered traffic per user (common to all regions) are shown in Table 3.11.

Offered traffic per cell

The parameter `Offered_Traffic_per_Cell`$_{e,s,l}$ [(calls · seconds)/(cell · hour)] gives the total traffic of service s offered in a cell of environment e on link l during the

Table 3.10   Number of users per cell.

Service	Environments		
	CBD	Urban pedestrian	Urban vehicular
Region 1			
S	514	22 759	1897
SM	281	12 471	1039
SD	91	4053	338
MMM	106	4677	390
HMM	106	4677	390
HIMM	176	7794	650
Region 2			
S	514	17 069	1897
SM	281	9353	1039
SD	91	3040	338
MMM	106	3507	390
HMM	106	3507	390
HIMM	176	5846	650
Region 3			
S	514	22 759	683
SM	281	12 471	374
SD	91	4053	122
MMM	106	4677	140
HMM	106	4677	140
HIMM	176	7794	234

Table 3.11   Offered traffic per user [(calls·seconds)/ (user·hour)].

Service	Link	Environments		
		CBD	Urban pedestrian	Urban vehicular
S		270.00	48.00	24.00
SM		2.80	0.90	0.60
SD		31.20	31.20	3.12
MMM	UL	4.28	3.42	0.07
	DL	22.50	18.00	0.36
HMM	UL	1.28	0.51	0.70
	DL	6.75	2.70	0.36
HIMM		16.80	8.40	1.32

busy hour. It is calculated by

$$\texttt{Offered\_Traffic\_per\_Cell}_{e,s,l} = \texttt{Offered\_Traffic\_per\_User}_{e,s,l}$$
$$\times \texttt{Users\_per\_Cell}_{e,s}. \qquad (3.8)$$

The value of `Offered_Traffic_per_Cell`$_{e,s,l}$ represents how many seconds per hour is used by the traffic from all users of service s in a cell of environment e on link l. As `Users_per_Cell`$_{e,s}$ depends on the region, so does `Offered_Traffic_per_Cell`$_{e,s,l}$.

The calculated numerical values of the offered traffic per cell in Region 1 are shown in Table 3.12. The corresponding values for Regions 2 and 3 can be calculated similarly.

Table 3.12　Offered traffic per cell [(calls·seconds)/(cell·hour)] in Region 1.

Service	Link	Environments		
		CBD	Urban pedestrian	Urban vehicular
S		1.39E+05[a]	1.09E+05	4.55E+04
SM		5.07E+02	1.12E+04	6.24E+02
SD		2.85E+03	1.26E+05	1.05E+03
MMM	UL	4.51E+02	1.60E+04	2.67E+01
	DL	2.38E+03	8.42E+04	1.40E+02
HMM	UL	1.35E+02	2.40E+03	2.67E+01
	DL	7.13E+02	1.26E+04	1.40E+02
HIMM		2.96E+03	6.55E+04	8.57E+02

[a]For example, '1.39E+05' means 1.39×10^5.

Offered traffic per group

The parameter `Offered_Traffic_per_Group`$_{e,s,l}$ (Erlang/group) gives the total traffic of service s offered in a group of cells of environment e on link l during the busy hour. It is calculated by

$$\texttt{Offered\_Traffic\_per\_Group}_{e,s,l} = \frac{\texttt{Offered\_Traffic\_per\_Cell}_{e,s,l}}{3600}$$
$$\times \texttt{Group\_Size}_{e,s}. \tag{3.9}$$

At this moment, we have converted the unit of offered traffic from [(calls · seconds)/ hour] to 'Erlang' by

$$1 \text{ Erlang} = 3600 \, \frac{\text{calls} \cdot \text{seconds}}{\text{hour}}, \tag{3.10}$$

which is physically dimensionless. This unit of traffic volume is named after a Danish mathematician Agner Krarup Erlang (1878–1929). The value of traffic volume in Erlang shows the proportion of the time that the communication channel is busy to transmit that traffic.[2]

The calculated numerical values of the offered traffic per group in Region 1 are shown in Table 3.13. The corresponding values for Regions 2 and 3 can be calculated similarly.

---

[2]See Fortet and Grandjean (1964) for the life and works of A. K. Erlang.

Table 3.13   Offered traffic per group (Erlang/group) in Region 1.

Service	Link	CBD	Urban pedestrian	Urban vehicular
			Environments	
S		269.64	2124.19	88.51
SM		0.99	21.82	1.21
SD		5.55	245.88	2.05
MMM	UL	0.88	31.10	0.05
	DL	4.62	163.68	0.27
HMM	UL	0.26	4.66	0.05
	DL	1.39	24.55	0.27
HIMM		5.75	127.31	1.67

3.3.2   Erlang-B and Erlang-C formulas

We can now determine the number of service channels required to support the traffic demand so as to satisfy the QoS requirement by the user. The QoS of circuit-switched services (S, SD and HIMM) is specified by the probability that a call is blocked because there are no channels available when it arises. The QoS of packet-switched services (SM, MMM and HMM) is specified by the probability that the waiting time of a session exceeds 50% of the average session duration. In order to find the minimum number of service channels that satisfies the user's requirement on these probabilities, we use the formulas of queueing theory, namely, Erlang-B formula for the circuit-switched services and Erlang-C formula for the packet-switched services.

In this section we present these formulas, show numerical values, and discuss some of their properties. In the next section we apply the formulas to the determination of required capacity.

Erlang-B formula for circuit-switched services

For circuit-switched services (S, SD and HIMM), we use a loss system model denoted by M/M/s/s in queueing theory, where s is the number of servers (channels) as well as the maximum number of calls that can be accommodated in the system (there is no waiting room).[3] For the offered traffic a, the probability that an arriving call is blocked (*blocking probability*) because all the servers are used when it arrives is given by the so-called *Erlang-B formula*:

$$E_B(s, a) := \frac{a^s/s!}{\sum_{k=0}^{s} a^k/k!}.$$ (3.11)

The derivation of the Erlang-B formula is given in Appendix A.1.

Figure 3.3 plots the blocking probability $E_B(s, a)$ against the offered traffic a for various values of the number s of servers ($s = 1$–15). We observe that:

- for given s, $E_B(s, a)$ increases monotonously as a increases: $\lim_{a \to \infty} E_B(s, a) = 1$;

- for given a, $E_B(s, a)$ decreases monotonously as s increases.

---

[3]See Appendix A for the shorthand notation of queueing models.

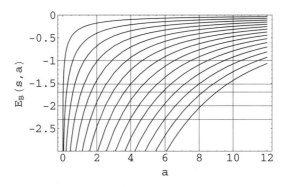

Figure 3.3 Blocking probability $E_B(s, a)$ against offered traffic a for an M/M/s/s loss system ($s = 1$–15 from left to right).

Table 3.14 shows the offered traffic a for the given values of s and $E_B(s, a)$ calculated by the Erlang B-formula. Furthermore, Table 3.15 shows the offered traffic a for the given values of $s = 16$–115 and $E_B(s, a) = 0.01$ for the convenience of the reader.

Table 3.14   Offered traffic a for given values of s and $E_B(s, a)$.

s	\multicolumn{6}{c}{$E_B(s, a)$}
	0.001	0.005	0.01	0.02	0.03	0.1
1	0.0010	0.0050	0.0101	0.0204	0.0309	0.1111
2	0.0458	0.1054	0.1526	0.2235	0.2816	0.5954
3	0.1938	0.3490	0.4555	0.6022	0.7151	1.2708
4	0.4393	0.7012	0.8694	1.0923	1.2589	2.0454
5	0.7621	1.1320	1.3608	1.6571	1.8252	2.8811
6	1.1459	1.6218	1.9090	2.2759	2.5431	3.7584
7	1.5786	2.1575	2.5009	2.9354	3.2497	4.6662
8	2.0513	2.7299	3.1276	3.6271	3.9865	5.5971
9	2.5575	3.3326	3.7825	4.3447	4.7479	6.5464
10	3.0920	3.9607	4.4612	5.0840	5.5294	7.5106
11	3.6511	4.6104	5.1599	5.8415	6.3280	8.4871
12	4.2314	5.2789	5.8760	6.6147	7.1410	9.4740
13	4.8305	5.9638	6.6072	7.4015	7.9667	10.4699
14	5.4464	6.6632	7.3517	8.2003	8.8035	11.4735
15	6.0772	7.3755	8.1080	9.0096	9.6500	12.4838

Erlang-C formula for packet-switched services

For packet-switched services (SM, MMM and HMM), we use a delay system model denoted by M/M/s in queueing theory, where s is the number of servers (channels). In the delay

Table 3.15   Offered traffic a for given values of s and $E_B(s, a) = 0.01$.

s	a	s	a	s	a	s	a
16	8.8750	41	29.8882	66	52.4353	91	75.6198
17	9.6516	42	30.7712	67	53.3531	92	76.5560
18	10.4369	43	31.6561	68	54.2718	93	77.4926
19	11.2301	44	32.5430	69	55.1915	94	78.4298
20	12.0306	45	33.4317	70	56.1120	95	79.3676
21	12.8378	46	34.3223	71	57.0335	96	80.3059
22	13.6513	47	35.2146	72	57.9558	97	91.2447
23	14.4705	48	36.1986	73	58.8789	98	82.1840
24	15.2950	49	37.0042	74	59.8028	99	83.1238
25	16.1246	50	37.9014	75	60.7276	100	84.0642
26	16.9588	51	38.8001	76	61.6531	101	85.0050
27	17.7974	52	39.7003	77	62.5794	102	85.9463
28	18.6402	53	40.6019	78	63.5065	103	86.8880
29	19.4869	54	41.5049	79	64.4343	104	87.8303
30	20.3373	55	42.4092	80	65.3628	105	88.7729
31	21.1912	56	43.3149	81	66.2920	106	89.7161
32	22.0483	57	44.2218	82	67.2219	107	90.6597
33	22.9087	58	45.1299	83	68.1524	108	91.6037
34	23.7720	59	46.0392	84	69.0837	109	92.5481
35	24.6381	60	46.9497	85	70.0156	110	93.4930
36	25.5070	61	47.8613	86	70.9481	111	93.4930
37	26.3785	62	48.7740	87	71.8812	112	95.3840
38	27.2525	63	49.6878	88	72.8150	113	96.3301
39	28.1288	64	50.6026	89	73.7494	114	97.2766
40	29.0074	65	51.5185	90	74.6843	115	98.2235

system, the waiting room is limitless. For the offered traffic a, we assume that $a < s$ for the stability, i.e. the condition that the queue of waiting sessions does not grow infinitely. Then the probability that the waiting time W of a session exceeds x is given by

$$P\{W > x\} = E_C(s, a)\, e^{-(s-a)\mu x} \quad x \geq 0, \tag{3.12}$$

where $1/\mu$ is the average service time. The factor

$$E_C(s, a) := \frac{a^s/(s-1)!(s-a)}{\sum_{k=0}^{s-1} a^k/k! + a^s/(s-1)!(s-a)} \tag{3.13}$$

is the probability $P\{W > 0\}$ that an arriving session waits because all servers are used when it arrives. Equation (3.13) is called the *Erlang-C formula*. The derivation of (3.12) and (3.13) is given in Appendix A.2.

Figure 3.4 plots the probability $P\{W > 0.5/\mu\}$ against the offered load a for various values of the number s of servers ($s = 1$–15). We have chosen the case $x = 0.5/\mu$ for plotting because this corresponds to the QoS requirement adopted in Report ITU-R M.2023. We may observe that:

- for given s, $P\{W > 0.5/\mu\}$ increases monotonously as a increases;

- for given a, $P\{W > 0.5/\mu\}$ decreases monotonously as s increases.

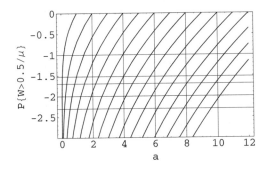

Figure 3.4 Probability $P\{W > 0.5/\mu\}$ against offered traffic a for an M/M/s delay system ($s = 1$–15 from left to right).

Table 3.16 shows the offered traffic a for the given values of s and $P\{W > 0.5/\mu\}$ calculated by (3.12) and (3.13). Furthermore, Table 3.17 shows the offered traffic a for the given values of $s = 16$–115 and $P\{W > 0.5/\mu\} = 0.01$ for the convenience of the reader.

Table 3.16   Offered traffic a for given values of s and $P\{W > 0.5/\mu\}$.

s	\multicolumn{6}{c}{$P\{W > 0.5/\mu\}$}
	0.001	0.005	0.01	0.02	0.03	0.1
1	0.0016	0.0082	0.0164	0.0324	0.0483	0.1527
2	0.0737	0.1646	0.2324	0.3278	0.4003	0.7183
3	0.3041	0.5246	0.6638	0.8400	0.9639	1.4451
4	0.6742	1.0247	1.2288	1.4741	1.6399	2.2469
5	1.1488	1.6193	1.8799	2.1838	2.3841	3.0922
6	1.7016	2.2805	2.5905	2.9445	3.1741	3.9659
7	2.3144	2.9910	3.3445	3.7421	3.9968	4.8597
8	2.9743	3.7393	4.1315	4.5673	4.8440	5.7683
9	3.6722	4.5174	4.9443	5.4140	5.7101	6.6884
10	4.4012	5.3196	5.7777	6.2780	6.5913	7.6176
11	5.1561	6.1417	6.6281	7.1560	7.4849	8.5541
12	5.9239	6.9803	7.4927	8.0456	8.3886	9.4969
13	6.7283	7.8330	8.3692	8.9450	9.3010	10.4444
14	7.5399	8.6977	9.2559	9.8528	10.2206	11.3965
15	8.3656	9.5729	10.1514	10.7679	11.1446	12.3523

Table 3.17   Offered traffic a for given values of s and $P\{W > 0.5/\mu\} = 0.01$.

s	a	s	a	s	a	s	a
16	11.0547	41	34.7758	66	59.2403	91	83.9236
17	11.9646	42	35.7467	67	60.2247	92	84.9135
18	12.8805	43	36.7816	68	61.2094	93	85.9035
19	13.8018	44	37.6913	69	62.1943	94	86.8936
20	14.7278	45	38.6648	70	63.1796	95	87.8839
21	15.6580	46	39.6391	71	64.1652	96	88.8743
22	16.5922	47	40.6140	72	65.1510	97	89.8649
23	17.5298	48	41.5897	73	66.1371	98	90.8555
24	18.4707	49	42.5660	74	67.1234	99	91.8464
25	19.4145	50	43.5429	75	68.1100	100	92.8373
26	20.3609	51	44.5204	76	69.0968	101	93.8284
27	21.3099	52	45.4985	77	70.0839	102	94.8196
28	22.2611	53	46.4771	78	71.0712	103	95.8109
29	23.2144	54	47.4563	79	72.0587	104	96.8023
30	24.1697	55	48.4360	80	73.0464	105	97.7938
31	25.1268	56	49.4161	81	74.0343	106	98.7855
32	26.0856	57	50.3967	82	75.0224	107	99.7772
33	27.0460	58	51.3778	83	76.0107	108	100.769
34	28.0079	59	52.3592	84	76.9992	109	101.761
35	28.9712	60	53.3411	85	77.9879	110	102.753
36	29.9358	61	54.3234	86	78.9768	111	103.745
37	30.9016	62	55.3061	87	79.9658	112	104.737
38	31.8685	63	56.2891	88	80.9550	113	105.730
39	32.8366	64	57.2725	89	81.9444	114	106.722
40	33.8507	65	58.2562	90	82.9339	115	107.715

3.3.3   Determination of required spectrum

We calculate the values of the following parameters sequentially in order to determine the required spectrum:

- Required_Channels_per_Group$_{e,s,l}$ (channels/group, dimensionless)
- Required_Channels_per_Cell$_{e,s,l}$ (channels/cell, dimensionless)
- Required_Bit_Rate_per_Cell$_{e,s,l}$ (Mbit/s/cell)
- Required_Spectrum$_{e,s,l}$ (MHz).

Required number of service channels per group

The number of channels for service s per group of cells in environment e on link l needed to satisfy the QoS requirement is denoted by Required_Channels_per_Group$_{e,s,l}$ (channels/group, dimensionless). For the given value of

$$a = \text{Offered\_Traffic\_per\_Group}_{e,s,l}, \tag{3.14}$$

we use the Erlang-B or Erlang-C formulas to determine the minimum value of

$$s = \texttt{Required\_Channels\_per\_Group}_{e,s,l} \qquad (3.15)$$

that satisfies the QoS requirement.

In Report ITU-R M.2023, the QoS requirement for circuit-switched service is specified in terms of the call-blocking probability. It is conceived that end users will expect the wireless QoS comparable to that of the fixed wireline network. Thus the QoS requirement is set to be the blocking probability of 1%, i.e.

$$E_B(s, a) < 0.01. \qquad (3.16)$$

For example, the offered traffic of SD service in the high-density in-building environment is 5.55 Erlang as shown in Table 3.13. We can then look at the case of $E_B(s, a) = 0.01$ either in Figure 3.14 or in Table 3.14 to know that 12 channels are sufficient. Thus we obtain the value 12 for the corresponding entry in Table 3.18.

The QoS requirement for packet-switched service is specified so that the probability that the waiting time of a session exceeds 50% of the average session duration be less than 1%, i.e.

$$P\{W > 0.5/\mu\} < 0.01. \qquad (3.17)$$

For example, the offered traffic of SM service in the urban pedestrian environment is 21.82 Erlang as shown in Table 3.18. We can then look at Table 3.17 to know that 28 channels are sufficient. Thus we obtain the value 28 for the corresponding entry in Table 3.18.

The required number of service channels per group in Region 1 obtained in this way is shown in Table 3.18.

Table 3.18   Required number of service channels per group in Region 1.

Service	Link	Environments		
		CBD	Urban pedestrian	Urban vehicular
S		293	2152	105
SM		4	28	4
SD		12	269	7
MMM	UL	4	38	2
	DL	9	171	3
HMM	UL	3	9	2
	DL	5	31	3
HIMM		12	146	6

Required number of service channels per cell

The required number of channels for service s per cell of environment e on link l is calculated by simply dividing the required number of service channels per group by the number of cells per group. It is denoted by `Required_Channels_per_Cell`$_{e,s,l}$ (channels/cell, dimensionless).

$$\texttt{Required\_Channels\_per\_Cell}_{e,s,l} = \frac{\texttt{Required\_Channels\_per\_Group}_{e,s,l}}{\texttt{Group\_Size}_{e,s}}.$$

(3.18)

The required number of service channels per cell in Region 1 is shown in Table 3.19.

Table 3.19   Required number of service channels per cell in Region 1.

Service	Link	Environments		
		CBD	Urban pedestrian	Urban vehicular
S		41.86	307.43	15.00
SM		0.57	4.00	0.57
SD		1.71	38.43	1.00
MMM	UL	0.57	5.43	0.29
	DL	1.29	24.57	0.43
HMM	UL	0.43	1.29	0.29
	DL	0.71	4.43	0.43
HIMM		1.71	20.86	0.86

Required bit rate per cell

Recall that the number of bits transmitted per unit time in a service channel is given by the service channel bit rate. Therefore, the required bit rate for service s per cell of environment e on link l can be calculated by multiplying the required number of service channels per cell with the service channel bit rate. It is denoted by `Required_Bit_Rate_per_Cell`$_{e,s,l}$ (Mbit/s/cell). Thus we obtain

$$\texttt{Required\_Bit\_Rate\_per\_Cell}_{e,s,l} = \texttt{Required\_Channels\_per\_Cell}_{e,s,l}$$
$$\times \frac{\texttt{Service\_Channel\_Bit\_Rate}_{e,s}}{1000}.$$

(3.19)

Here we have changed the unit of the channel bit rate from kbit/s to Mbit/s.

The calculated numerical values of the required bit rate per cell in Region 1 are shown in Table 3.20.

Required spectrum

The spectrum bandwidth in Hz needed to carry the data rate of 1 bit/s per cell is given by the net system capability `Net_System_Capability`$_{e,s}$ (bit/s/Hz/cell). Therefore,

Table 3.20 Required bit rate per cell (Mbit/s/cell) in Region 1.

			Environments	
Service	Link	CBD	Urban pedestrian	Urban vehicular
S		0.67	4.92	0.24
SM		0.01	0.06	0.008
SD		0.11	2.46	0.06
MMM	UL	0.04	0.35	0.02
	DL	0.49	9.44	0.16
HMM	UL	0.05	0.16	0.04
	DL	1.43	8.86	0.86
HIMM		0.22	2.67	0.11

the amount of spectrum required for service s for environment e on link l is calculated by dividing the required bit rate per cell by the net system capability. It is denoted by `Required_Spectrum`$_{e,s,l}$ (MHz). Thus we have

$$\mathtt{Required\_Spectrum}_{e,s,l} = \frac{\mathtt{Required\_Bit\_Rate\_per\_Cell}_{e,s,l}}{\mathtt{Net\_System\_Capability}_{e,s}}. \tag{3.20}$$

The calculated numerical values are shown in Table 3.21 for Region 1. The reader should be able to obtain similar tables for Regions 2 and 3.

Below we simply write $F_{e,s,l}$ for `Required_Spectrum`$_{e,s,l}$ for link direction $l \in$ {uplink, downlink}.

3.3.4 Weighting and adjustment

In the final stage of spectrum requirement calculation, we do some adjustment.

Weighting factor

The weighting factor $\alpha_{e,s}$ (dimensionless) provides appropriate weighting in the spectrum requirement calculations to take account of those factors that may be specific to a particular environment and/or service, and includes the following:

- weighting to adjust for geographical offsets in overlapping environments

- weighting to correct for non-simultaneous busy hour traffic environments.

The value for $\alpha_{e,s}$ may range between 0 and 1. In the Recommendation ITU-R M.1390, the default value is unity

$$\alpha_{e,s} = 1, \tag{3.21}$$

which is also assumed in Report ITU-R M.2023.

Let F_l (MHz) be the weighted sum of the required spectrum on each link over all environments and services:

$$F_l = \sum_e \sum_s \alpha_{e,s} F_{e,s,l} \quad l \in \{\text{uplink, downlink}\}. \tag{3.22}$$

Table 3.21   Required spectrum (MHz/cell) in Region 1.

Service	Environments			Total
	CBD	Urban pedestrian	Urban vehicular	
Uplink				
S	9.57	70.27	3.43	83.27
SM	0.06	0.45	0.06	0.58
SD	0.88	19.68	0.51	21.07
MMM	0.29	2.78	0.15	3.22
HMM	0.44	1.32	0.29	2.05
HIMM	1.76	21.36	0.88	23.99
Sum	13.00	115.85	5.32	134.16
Downlink				
S	9.57	70.27	3.43	83.27
SM	0.06	0.45	0.06	0.58
SD	0.88	19.68	0.51	21.07
MMM	3.95	75.48	1.32	80.75
HMM	11.43	70.86	6.86	89.14
HIMM	1.76	21.36	0.88	23.99
Sum	27.64	258.09	13.06	298.79

The total required spectrum on each link and their sums are shown separately for the three regions in the column F_l of Table 3.23.

If there is only a single network operator in the region, the calculation ends at this point.

Adjustment factor

The adjustment factor β (dimensionless), which is independent of any particular environment or service, accounts for the fact that additional spectrum will be needed if administrations decide to license multiple network operators. It provides for impacts as follows:

- Guard bands

  Guard bands have to be accommodated between two operators as well as at the boundaries between mobile and other services. The average value to compensate for this loss is estimated to be 4% uniformly. The increments of the spectrum by the guard band are shown in the column 'Guard band' in Table 3.23.

- Reduced trunking efficiency

  The adjustment due to the trunking inefficiency can be calculated based on the Erlang-B formula for the traffic demand of 80 Erlang. As an example, let us take the uplink of Region 2 which has six network operators with equal traffic share and identical services. If a total traffic demand of 80 Erlang is to be served in a single channel group (i.e. contiguous allocation of spectrum) with 1% blocking probability, it will require 96 channels (see Table 3.15). However, if the same 80 Erlang of traffic is divided equally among six different channel groups (corresponding to six networks), each network's group of channels will be required to carry $80/6 = 13.33$ Erlang of traffic. The Erlang-B formula for 13.33 Erlang with 1% blocking probability predicts that 22 channels must be required by each network, which implies that a total of 132 channels by all networks. This is an increase of $132/96 = 1.375$ over the number of channels needed by a single network to carry the same traffic. Thus we assume that 37.5% increase in spectrum is required for a region with six operators (Region 2).

Table 3.22 shows the increase in spectrum for various numbers of network operators in a region. Note that the amount of increase is not necessarily monotone in the number of networks.

According to this table, the 18.75% and 10.42% increase in spectrum is required in Region 1 with three networks and in Region 3 with two networks, respectively.[4] The increments of the spectrum by the reduced trunking efficiency are shown in the column 'Trunking' in Table 3.23. The spectrum requirements after considering the guard band and the reduced trunking efficiency are shown in the column 'Before M'.

Table 3.22   Impact of multiple networks on the trunking efficiency for 80 Erlang of traffic.

Number of networks	Traffic per network (Erlang)	Number of channels per network	Total number of channels	Ratio
1	80	96	96	1
2	40	53	106	1.1042
3	26.67	38	114	1.1875
4	20	30	120	1.25
5	16	25	125	1.3021
6	13.33	22	132	1.375
7	11.43	20	140	1.4583
8	10	18	144	1.5
9	8.89	17	153	1.5938
10	8	15	150	1.5625

- 5 MHz modularity

Let us again take the example of uplink in Region 2 with six operators, where 113.16 MHz of spectrum is required before considering the *modularity* rounding up. We first divide 113.16 MHz by 6 which is 18.86 MHz, and then round upward to the nearest multiple of 5 MHz to get 20 MHz for each network. Thus $20 \times 6 = 120$ MHz of spectrum is required in total after the modularity rounding up. A similar calculation shows that the 268.06 MHz of spectrum requirement before modularity rounding up results in 270 MHz on the downlink in Region 2. Therefore, a total of 390 MHz is

---

[4]In Table 15 of Report ITU-R M.2023, the increase in Region 3 with two networks is given as 7.00%, which seems to be incorrect. Accordingly, the numerical values for Region 3 in Table 3.23 differ from those in Report ITU-R M.2023.

required in Region 2. The spectrum requirements after modularity rounding up are shown in the column 'After M' in Table 3.23.

Table 3.23   Total spectrum requirement (MHz).

Region	Link	F_l	Guard band	Trunking	Before M	After M
1	UL	134.16	5.37	25.16	164.68	165
3 networks	DL	298.79	11.95	56.02	366.76	375
	Sum	432.95	17.32	81.18	531.45	540
2	UL	79.97	3.20	29.99	113.16	120
6 networks	DL	189.44	7.58	71.04	268.06	270
	Sum	269.41	10.78	101.03	381.22	390
3	UL	131.64	5.27	13.72	150.62	160
2 networks	DL	293.55	11.74	30.59	335.88	340
	Sum	425.19	17.01	44.30	486.50	490

Final total spectrum requirement

In Recommendation ITU-R M.1390, the total spectrum requirement (MHz) is expressed as

$$F = \beta \sum_e \sum_s \sum_l \alpha_{e,s} F_{e,s,l} \tag{3.23}$$

for each region. However, in the numerical examples given in Report ITU-R M.2023, the factor β does not seem to be a simple multiplicative parameter as in (3.23). Rather, the final total spectrum requirement F is calculated through a procedure as given in the above example.

3.4   Sequel to the Story

Table 3.24 shows a summary of global mobile terrestrial spectrum requirement copied from Table 1 in Report ITU-R M.2023, where each column shows:

A Forecast total terrestrial mobile spectrum requirement for the year 2010

B Identified total terrestrial mobile spectrum (including RR No. 5.388 IMT-2000 spectrum)

C Forecast additional IMT-2000 terrestrial component spectrum requirement for 2010.

Column A of Table 3.24 resulted from the calculation in Report ITU-R M.2023. The calculations in Report ITU-R M.2023 were produced by the MS Excel implementation of the calculation methodology which consists of one spreadsheet for each of the three ITU regions, as described in Tables 18–20 of Report ITU-R M.2023, with a limited set of macros performing the more complicated calculations. We note that there is minor discrepancy

Table 3.24 Summary of spectrum requirement (MHz).

Region	A	B	C
1	555	395	160
2	390	230	160
3	480	320	160

between our result shown in the rightmost column of Table 3.23 and the result in column A of Table 3.24. This is due to our correction made in the adjustment process. The spectrum in column B of Table 3.24 is the sum of the spectrum that already existed for 1G and 2G mobile systems and the spectrum requirement identified in the RR No. 5.388 for IMT-2000; see Figure 2.3. As a result, the additional spectrum requirement for the terrestrial component of IMT-2000 for the year 2010 was found to be 160 MHz in all regions as shown in column C of Table 3.24. This was used as input to the World Radiocommunication Conference 2000 (WRC-2000).

The results were included in the Conference Preparatory Meeting (CPM) Report to the WRC-2000 as a basis for making decision. The WRC-2000 identified additional spectrum bands for IMT-2000 as follows:

- 806–960 MHz

- 1710–1885 MHz

- 2500–2690 MHz.

The utilization of those bands was expected to be started sometimes after the year 2007, depending on the situation in particular countries.

The methodology described in Recommendation ITU-R M.1390 was later taken as a basis when the spectrum requirements for broadband nomadic wireless access (NWA) systems, including radio local area networks (RLANs), were calculated as part of the preparation for the World Radiocommunication Conference 2003 (WRC-03) Agenda Item 1.5 in Recommendation ITU-R M.1651. The traffic characteristics and the technical characteristics used in the calculation in Rec. ITU-R M.1651 were specific for RLANs, but the methodology was the same as that in Rec. ITU-R M.1390. As a result, the spectrum was identified in the 5 GHz range for RLANs at the WRC-03.

4

Spectrum Requirement Calculation for IMT-Advanced

**Marja Matinmikko, Jörg Huschke, Tim Irnich,
Naoto Matoba, Jussi Ojala, Pekka Ojanen,
Hideaki Takagi, Bernhard H. Walke and Hitoshi Yoshino**

The spectrum requirement calculation methodology for IMT-2000 described in Chapter 3 is not directly applicable to calculating the spectrum requirements for the future development of IMT-2000 and IMT-Advanced, which was the topic of World Radiocommunication Conference 2007 (WRC-07) held in Geneva in October–November 2007. Therefore in preparation for the WRC-07, ITU has developed a new methodology to calculate the spectrum requirements of the future development of IMT-2000 and IMT-Advanced, and presented it in Recommendation ITU-R M.1768.

This chapter introduces in detail the ITU spectrum calculation methodology for the future development of IMT-2000 and IMT-Advanced and highlights the most important steps in the methodology flow. The methodology follows a deterministic flow starting from the market predictions of the future mobile services and ending in the final spectrum requirements of pre-IMT, IMT-2000, future development of IMT-2000 and IMT-Advanced systems in the time span of years 2010–2020.

This chapter is organized as follows. Section 4.1 gives an overview of the development of the spectrum requirement calculation methodology and its relations to other ITU preparatory work for WRC-07. Section 4.2 introduces the models and input parameters used in the methodology. Section 4.3 presents the actual calculation algorithms used in the methodology including the traffic demand calculations and distributions, the capacity calculations and the spectrum calculations. Finally, Section 4.4 provides a summary of the methodology by revisiting the relations of the different definitions used in the methodology.

Spectrum Requirement Planning in Wireless Communications   Edited by Hideaki Takagi and Bernhard H. Walke
© 2008 John Wiley & Sons, Ltd

A comprehensive study of the methodology, calculation tool, input market parameters and radio-related parameters (corresponding to Chapters 4, 5, 6 and 7, respectively, of this book) and their influence on the spectrum requirements is given by Matinmikko (2007).

4.1   Overview

The development of the spectrum calculation methodology for the future development of IMT-2000 and IMT-Advanced was carried out in the ITU during 2003–2005. Starting from the assessment of the previous ITU methodology for IMT-2000 as described in Section 4.1.1, the development of the new ITU methodology involved different stages as described in Section 4.1.2. Along with the development of the spectrum calculation methodology, the ITU also conducted other studies related to the spectrum requirements of the future development of IMT-2000 and IMT-Advanced. The relations of the methodology to other ITU preparations for WRC-07 are summarized in Section 4.1.3. The deterministic flow chart of the spectrum calculation methodology used as the framework for the development of the methodology is presented in Section 4.1.4.

4.1.1   Limitation of methodology for IMT-2000

The ITU methodology for estimation of spectrum requirements for IMT-2000 systems presented in Recommendation ITU-R M.1390 and summarized in Chapter 3 is based on blended 2G and IMT-2000 networks. The methodology follows a straightforward deterministic approach and its implementation in MS Excel includes only a single worksheet per region. Therefore, the overall methodology is not very complex and the calculations can be performed very quickly.

The methodology in Rec. ITU-R M.1390 considers a single network with service delivery based on circuit-switching. However, according to the framework and objectives for IMT-Advanced shown in Recommendation ITU-R M.1645, the service delivery in the future is mainly based on packet switching by the Internet Protocol (IP), and the seamless interworking between different access systems is required. The spectrum calculation methodology for IMT-2000 does not, therefore, characterize the operating environment of IMT-Advanced adequately. The methodology in Rec. ITU-R M.1390 was developed to handle both circuit-switched and packet-switched data but the approach was based on circuit-switching, while the packet-switching was not modeled properly. Instead, the spectrum calculation methodology for IMT-Advanced needs to model packet-switching more accurately and accommodate multiple networks supporting rather dissimilar service structures.

In Rec. ITU-R M.1390, the services were characterized with the net user bit rate in kbit/s, which was assumed to be constant and identical in all environments. In the future, services will have different characteristics with respect to the bit rate, delay and other Quality of Service (QoS) requirements differing according to the environments and usage scenarios. In IMT-2000 systems, the services were voice based, including short message service (SMS) and some higher data rate services. Future services will be more diverse and thus the methodology described in Chapter 3 does not model them adequately.

4.1.2   Development of methodology for IMT-Advanced

Due to the limitations in modeling future services, networks and operating environment, the spectrum calculation methodology for IMT-2000 presented in Rec. ITU-R M.1390 is not directly applicable to estimate the spectrum requirements of IMT-Advanced systems. Therefore, ITU has developed a new methodology to calculate the spectrum requirements of the terrestrial component of the future development of IMT-2000 and IMT-Advanced for the WRC-07. First contributions towards a new methodology were made by the Mobile IP-based Network Developments project (MIND) funded by the European Commission Information Society Technologies (IST) Framework Programme 5 (FP5) in 2002 (IST-MIND D3.3 2002). However, the methodology presented there did not meet with success, and new proposals for the development of the methodology were received at Working Party 8F of ITU-R (ITU-R WP 8F) in 2003–2005.

Continuation for the ITU methodology for estimation of spectrum requirements of IMT-2000 was presented by Mohr (2003), where the impact of estimated data rate requirements on the spectrum requirements of future systems was examined. In Mohr (2003), the radio interface is described in a generic form based on a modified Shannon channel capacity equation, which is useful to obtain generalized results independent of the particular radio interface. By defining a relation between cluster size, carrier-to-interference ratio and coverage requirements, the spectrum requirement and system throughput can be calculated for a radio interface with and without adaptive modulation and coding.

The new spectrum calculation methodology, accepted by ITU, for the future development of IMT-2000 and IMT-Advanced which goes beyond IST-MIND D3.3 (2002) and Mohr (2003) is presented in IST-2003-50781 WINNER D6.2 (2005). The work on the spectrum calculation methodology by ITU-R WP 8F was completed in November 2005 when ITU-R Study Group (SG) 8 adopted the final Recommendation ITU-R M.1768. Before completion, the development of the methodology had continued within the ITU-R WP 8F for several years.

The first stage in the development of the spectrum calculation methodology in ITU-R WP 8F was to set the guidelines for the development of the methodology. The guidelines were agreed in 2003. In the following stage, the methodology flow chart shown in Figure 4.2 was agreed in 2004. This flow chart presents the general structure of the methodology, including the different processing steps. After the main structure of the methodology was agreed, the individual processing steps in the methodology were developed by following the agreed guidelines and flow chart. The major parts of the methodology were developed in 2004, and agreed in the beginning of 2005.

After the major parts of the spectrum calculation methodology were developed, the implementation of the methodology into a software tool was started in 2005. Along with the implementation work, the spectrum calculation methodology was refined and the details of the processing steps were specified. In particular, the different input parameters to the methodology were discussed. Finally, the methodology document Rec. ITU-R M.1768 was agreed in ITU-R WP 8F in October 2005, and then adopted by ITU-R SG 8 in November 2005. The recommendation was published in March 2006. The first version of the software tool implementing the whole methodology was released for ITU-R active members in November 2005. The tool was later revised, and the final version, including the actual values for the input parameters, was made available in May 2006. This tool, implemented in

MS Excel and now publicly available on the ITU web site 'http://www.itu.int/ITU-R/study-groups/docs/speculator.doc', is introduced in Chapter 5 of this book.

4.1.3   ITU preparation for WRC-07

In preparation for WRC-07 on the frequency-related matters of the future development of IMT-2000 and IMT-Advanced, ITU has published the following documents:

- Recommendation ITU-R M.1768, Methodology for calculation of spectrum requirements for the future development of the terrestrial component of IMT-2000 and systems beyond IMT-2000;

- Report ITU-R M.2072, World mobile telecommunication market forecast;

- Report ITU-R M.2074, Radio aspects for the terrestrial component of IMT-2000 and systems beyond IMT-2000;

- Report ITU-R M.2078, Estimated spectrum bandwidth requirements for the future development of IMT-2000 and IMT-Advanced;

- Report ITU-R M.2079, Technical and operational information for identifying spectrum for the terrestrial component of future development of IMT-2000 and IMT-Advanced.

For the spectrum requirement calculations, the methodology is implemented in a calculation tool available in

'SPECULATOR' – Tool for estimating the spectrum requirements for the future development of IMT-2000 and IMT-Advanced.
http://www.itu.int/ITU-R/study-groups/docs/speculator.doc

This is the implementation in MS Excel of the ITU-accepted spectrum calculation methodology shown in Rec. ITU-R M.1768. The results of using the tool to estimate the spectrum requirements are collected in Report ITU-R M.2078, which also includes all the input parameter values for the calculations. The input to the calculations includes market studies from Report ITU-R M.2072 as well as radio parameters from Report ITU-R M.2074. Possible candidate bands to fulfill the estimated spectrum demand are discussed in Report ITU-R M.2079. Finally, all contributions are combined in the Conference Preparatory Meeting (CPM) Report to WRC-07

ITU-R CPM07-2. Report of the CPM to WRC-07, 2007.
http://www.itu.int/md/R07-CPM-R-0001/en

The relationship between these documents are illustrated in Figure 4.1.

4.1.4   Flow chart of methodology for IMT-Advanced

The ITU spectrum calculation methodology for the future development of IMT-2000 and IMT-Advanced considers the whole market of wireless services and models both reservation-based and packet-based service deliveries. The generic flow chart of the methodology is shown in Figure 4.2.

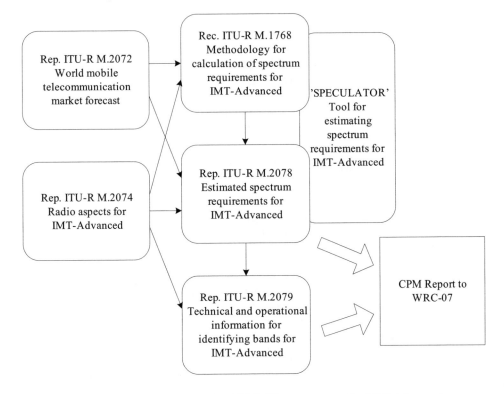

Figure 4.1 ITU preparations for WRC-07 on spectrum for IMT-Advanced.

The methodology follows a deterministic approach and consists of nine steps starting from the market expectations of wireless communication services and ending in the final spectrum requirements of the family of IMT systems. Step 1 introduces the definitions of the different concepts used in the methodology, as well as the parameters used to characterize the concepts. The definitions for service categories, service environments, radio environments and radio access technique groups, and the associated parameters are explained in Sections 4.2.1–4.2.4.

The second step in the methodology flow chart is to analyze the market data which is obtained from the market studies given in Report ITU-R M.2072. Step 3 is to process the market studies further to obtain the traffic values for the service categories in different environments needed in the spectrum calculation methodology. The fourth step distributes the total traffic of all mobile communication systems to different radio environments (i.e. cell layers) and radio access techniques (RATs) which are presented as RAT groups (RATGs). Steps 2–4 are described in Sections 4.3.1–4.3.3. Four RATGs are considered up to Step 4, but from Step 5 onwards the calculations are done for only RATG 1 and RATG 2 which correspond to pre-IMT, IMT-2000, future development of IMT-2000, and IMT-Advanced.

Step 5 is to determine the required system capacity to carry the offered traffic. The methodology uses separate capacity calculation algorithms for reservation-based and packet-based services, which are presented in Sections 4.3.4 and 4.3.5, respectively. Step 6 calculates

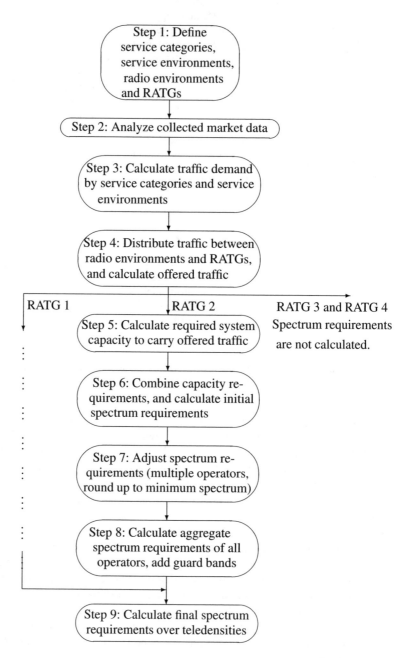

Figure 4.2 Flow chart of ITU spectrum calculation methodology for IMT-Advanced.

the spectrum requirements of two RATGs. Step 7 is to apply necessary adjustments for the spectrum results to take into account practical network deployments and to aggregate over the different spatially coexisting cell layers. The aggregate spectrum requirements of all operators are calculated in Step 8. Step 9 is to obtain the final spectrum requirements of the pre-IMT, IMT-2000, future development of IMT-2000 and IMT-Advanced by taking the maximum over spatially non-coexisting environments. Steps 6–9 are described in Section 4.3.6.

The flow chart in Figure 4.2 of the ITU spectrum calculation methodology for the future development of IMT-2000 and IMT-Advanced is similar to the flow chart in Figure 3.1 of the ITU spectrum calculation methodology for IMT-2000. Both methodologies follow a deterministic approach and include similar processing steps. However, the ITU spectrum calculation methodology for the future development of IMT-2000 and IMT-Advanced is more advanced, and includes features which were not included in the methodology for IMT-2000. The ITU spectrum calculation methodology for the future development of IMT-2000 and IMT-Advanced considers the whole market of wireless services, and uses comprehensive market studies to predict the traffic volumes in different cell layers of the wireless systems. In particular, the new ITU methodology employs advanced methods of calculating the capacity requirements of the future services and models the packet-based services more accurately.

4.2 Models and Input Parameters

The first step in the spectrum requirement calculation methodology flow chart in Figure 4.2 includes the definition of the different concepts used in the methodology to model the real situations. Service categories, service environments, radio environments and RATGs are defined. Each definition may contain several parameters to characterize the definitions which are needed as input in the calculations. The following sections briefly explain the different models used to characterize the environments, services and radio systems together with their associated input parameters. The input parameters used in the different steps of the methodology are summarized in Figure 4.3.

4.2.1 Service categories

A *service category* (SC) is a combination of a service type and an associated traffic class. *Service types* are characterized by the service bit rates. Five service types are considered, including very low rate data, low rate data and low multimedia, medium multimedia, high multimedia and super-high multimedia. *Traffic classes* are characterized by their sensitivity to delay. Four traffic classes are considered, including conversational, streaming, interactive and background. The traffic in conversational and streaming classes must be delivered with low delay (real-time) and low variation in delay (jitter). Traffic in interactive and background classes may be delivered with *best efforts* meaning that the network does not guarantee the timely delivery; however, the payload contents must be preserved; see Table 4.1. The combinations of service types and traffic classes lead to the 20 SCs considered in the methodology as shown in Table 4.2. Index n is used to denote the different service categories.

Service categories are characterized with parameters which are obtained either from market studies or from other sources. The following parameters are obtained from the ITU market study given in Report ITU-R M.2072 (the first column in Figure 4.3):

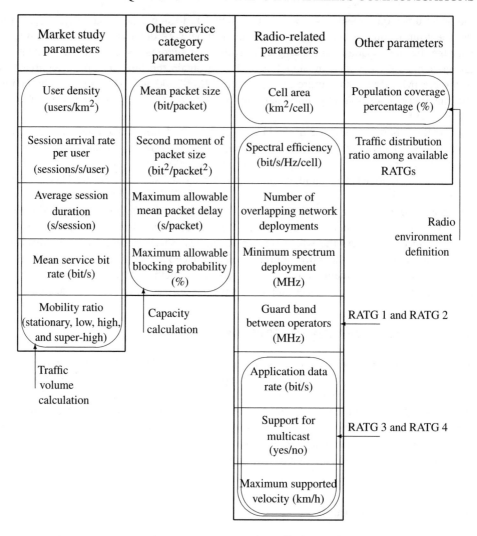

Figure 4.3 Input parameters for ITU spectrum calculation methodology.

- user density (users/km^2)

- session arrival rate per user (sessions/s/user)

- average session duration (s/session)

- mean service bit rate (bit/s)

- mobility ratios (%, dimensionless).

Numerical examples of these market study parameter values are given in Appendix B.

Table 4.1   Characteristic and example applications of traffic classes.

Traffic class	Conversational (real-time)	Streaming (real-time)	Interactive (best efforts)	Background (best efforts)
Characteristics	Stringent and low delay	Preserve delay variation	Preserve payload contents	Preserve payload contents
	Preserve delay variation		Request response pattern	Data is not expected within a certain time
Example applications	Voice	Streaming video	Web browsing	Download of e-mails

Table 4.2   Service categorization.

	Traffic classes			
Service type	Conversational	Streaming	Interactive	Background
Super-high multimedia	SC 1	SC 6	SC 11	SC 16
High multimedia	SC 2	SC 7	SC 12	SC 17
Medium multimedia	SC 3	SC 8	SC 13	SC 18
Low rate data/low multimedia	SC 4	SC 9	SC 14	SC 19
Very low rate data	SC 5	SC 10	SC 15	SC 20

The market studies characterize four mobility ratios as proportions of users for such classes as stationary, low, high, and super-high. The spectrum calculation methodology considers the following three mobility ratios and their associated velocity limits:

- stationary/pedestrian (0–4 km/h)

- low ($>$ 4 km/h and $<$ 50 km/h)

- high ($>$ 50 km/h).

The market study mobility ratios obtained for stationary, low, high and super-high mobility need to be mapped into the methodology mobility ratios for stationary/pedestrian, low and high mobility. The mapping is done using J_m factors, for service environment m, which are coefficients between 0 and 1. They are used to divide the high and super-high mobility ratios from market studies between the low and high mobility ratios used in the methodology; see Sections 4.3.1 and 6.3.6 for details.

In addition to the market-related service category parameters, the methodology also requires other parameters to characterize the service categories, including (the second column in Figure 4.3):

- mean packet size (bit/packet)

- second moment of packet size ($bit^2/packet^2$)

- maximum allowable mean packet delay (s/packet)

- maximum allowable blocking probability (%).

Service categories are delivered using either a reservation-based (circuit switched) or a packet-based transmission scheme. The conversational and streaming traffic classes (SC 1–10) are assumed to be served with circuit switching, while interactive and background traffic classes (SC 11–20) are served with packet switching.

A service category can include unicast and multicast traffic, which are treated separately in the methodology. Unicast traffic is transmitted from a transmitter to a single specified receiver, whereas multicast traffic is transmitted to a group of multiple specified receivers simultaneously.

4.2.2   Service environments

Service environments (SEs) are used to characterize the different environments where the users exist. A service environment is a combination of teledensity and service usage pattern. *Teledensities* describe the population density in different areas. Three teledensities are considered, including dense urban, suburban and rural. Index d is used to denote the different teledensities:

$$d \in \{\text{dense urban, suburban, rural}\}. \tag{4.1}$$

Service usage patterns describe the user behavior. Three service usage patterns are considered, including home, office and public area.

Teledensities are geographically non-overlapping areas while several service usage patterns can coexist in each teledensity, resulting in possibly several service environments in each teledensity. Market data is given per service environment while the spectrum requirements are calculated per teledensity. The methodology considers six service environments, SE 1–6, as shown in Table 4.3. Index m is used to denote the different service environments:

$$m \in \{1, \ 2, \ 3, \ 4, \ 5, \ 6\}. \tag{4.2}$$

Table 4.3   Identification of service environments.

Service usage pattern	Teledensities		
	Dense urban	Suburban	Rural
Home	SE 1	SE 4	SE 6
Office	SE 2	SE 5	SE 6
Public area	SE 3	SE 5	SE 6

4.2.3   Radio environments

Radio environments (REs) are defined by the cell layers in a network consisting of hierarchical cell layers, i.e. macro cell, micro cell, pico cell and hot spot. Radio environments are the areas exhibiting common propagation and deployment conditions. Index p is used to denote the different radio environments:

$$p \in \{\text{macro cell, micro cell, pico cell, hot spot}\}. \quad (4.3)$$

The radio environments are characterized by two parameters needed in the calculations (the first row in Figure 4.3):

- cell area $(\text{km}^2/\text{cell})$

- population coverage percentage (%).

The *cell area* of a radio environment may vary depending on the teledensity. The cell area of radio environment p in teledensity d is denoted by $A_{d,p}$. An example of the cell areas of different radio environments in different teledensities are given in Table 7.1.

The availability of radio environments depends on the service environment. It is possible that some radio environments are supported in all service environments. In practice, the total area of a service environment is covered by a certain radio environment up to only a certain percentage, which is defined as the *population coverage percentage*. The population coverage percentage of different radio environments in different service environments denotes the ratio of population that lives in the service area of the given radio environment in the given service environment. The population coverage percentages of the macro cell, micro cell, pico cell and hot spot radio environments are denoted by X_{macro}, X_{micro}, X_{pico} and X_{hs}, respectively. We note that $X_{\text{macro}} = 100\%$ always, because a macro cell should cover all the population. Zero population coverage percentage implies that the given radio environment does not exist in the given service environment.

Examples of the population coverage percentages of different radio environments in different service environments are given in Tables 7.8–7.10 for different years. According to Table 7.8, the population coverage percentages in service environment 1 (home in the dense urban teledensity) in the forecast year 2010 are:

$$X_{\text{macro}} = 100\%, \quad X_{\text{micro}} = 90\%, \quad X_{\text{pico}} = 0, \quad X_{\text{hs}} = 80\%.$$

Thus in this service environment, all population is in the macro cell, 90% of population in the micro cell, 80% of population in the hot spot, and none in the pico cell (the pico cells are not deployed).

4.2.4 Radio access technique groups

The ITU spectrum calculation methodology takes into account the total mobile telecommunication market provided by various communication means. The methodology needs to be technology neutral and generic, and therefore the individual radio access techniques (RATs) are grouped into four RATGs. The grouping is presented in Report ITU-R M.2074 and the four RATGs are:

- RATG 1: Pre-IMT systems, IMT-2000, and its future development;

- RATG 2: IMT-Advanced, e.g. new mobile access and new nomadic/local area wireless access;

- RATG 3: Existing radio local area networks (RLANs) and their enhancements;

- RATG 4: Digital mobile broadcasting systems and their enhancements.

All four RATGs are considered up to Step 4 in the methodology flow chart in Figure 4.2, i.e. up to and including the distribution of traffic. From Step 5 onwards, only RATG 1 and RATG 2 are considered. Finally, the spectrum requirements are calculated for only RATG 1 and RATG 2, which correspond to pre-IMT, IMT-2000 and its future development, and IMT-Advanced. Index *rat* is used to denote the different RATGs.

RATGs are characterized by radio-related parameters (the third column in Figure 4.3). The following radio-related parameters are required for all RATGs:

- application data rate (bit/s);

- maximum supported velocity (stationary/pedestrian, low, high);

- support for multicast (yes/no).

For RATG 1 and RATG 2, the methodology also requires values for the following radio-related parameters:

- spectral efficiency for unicast and multicast traffic (bit/s/Hz/cell);

- number of overlapping network deployments;

- minimum spectrum deployment per operator per radio environment (MHz);

- guard band between operators (MHz).

Numerical examples of these radio-related parameters are given in Tables 7.2–7.7. The ranges of values for radio-related parameters developed by ITU-R WP 8F are given in Report ITU-R M.2074. The values agreed by ITU and eventually used in the calculations are given in Report ITU-R M.2078. The spectral efficiency is an important parameter in the calculations as it is used to transform the capacity requirements into spectrum requirements.

In addition to the radio-related parameters listed above, the methodology also requires distribution ratios among the available RATGs (ξ_{rat} in Equation (4.10) below) as input for the calculations (the fourth column in Figure 4.3). This parameter gives the fractions according to which the offered traffic is distributed among the different RATGs in a given radio environment. Numerical examples of distribution ratios among available RATGs are given in Tables 7.11–7.13.

4.3   Methodology

The models for environments, services and radio systems and their associated parameters described in Section 4.2 are used in the actual calculation methodology, which is explained in detail in the following subsections. The methodology includes traffic calculation and distribution, capacity requirement calculation and finally spectrum requirement calculation.

The treatment in this section follows Recommendation ITU-R M.1768 with the exception that the time-dependency and the flexible spectrum use (FSU) for describing temporally and regionally varying natures of traffic are omitted, because they are not used in the actual calculations performed in Report ITU-R M.2078.

4.3.1   Calculation of traffic demand from market data

The second and third steps of the methodology flow chart in Figure 4.2 are to analyze the collected market data and to calculate the traffic demand by service categories and service environments. The market data includes the traffic forecasts covering the whole market of mobile telecommunications in the years 2010, 2015 and 2020. The market data is needed for the following parameters: user density, session arrival rate per user, average session duration, mean service bit rate, and mobility ratios for different service categories in different service environments in different forecast years. ITU-R provides a comprehensive collection of market studies in Report ITU-R M.2072. The data from the market report is further processed to obtain the input traffic parameter values for the methodology, which corresponds to Step 3. As the output from Steps 2 and 3, the market data for the five parameters is given per service category and service environment in three different forecast years 2010, 2015 and 2020 separately for uplink and downlink directions as well as for unicast and multicast traffic.

The market studies in Report ITU-R M.2072 define the mobility classes differently compared to those used in the methodology. Therefore, the market study parameter values are further processed to obtain the input parameters to the methodology. The market study mobility ratios are split to the methodology mobility ratios by using the splitting factor J_m for service environment m. The specific equations for the processing of market data are given in Section 6.3.6.

The market study traffic volumes per service are mapped into the following methodology traffic demand per service category:[1]

- $U_{m,n}$: user density of service category n in service environment m (users/km^2);

- $Q_{m,n}$: session arrival rate per user of service category n in service environment m (sessions/s/user);

- $\mu_{m,n}$: average session duration of service category n in service environment m (s/session);

- $r_{m,n}$: mean service bit rate of service category n in service environment m (bit/s).

The market study mobility ratios for stationary, low, high and super-high mobilities are mapped into the following methodology mobility ratios for stationary/pedestrian, low and high mobility classes:

- $MR_{stat/ped;m,n}$: stationary/pedestrian mobility ratio of service category n in service environment m;

- $MR_{low;m,n}$: low mobility ratio of service category n in service environment m;

- $MR_{high;m,n}$: high mobility ratio of service category n in service environment m.

Note that mobility ratios are dimensionless taking values between 0 and 1 and that they should sum to one:

$$MR_{stat/ped;m,n} + MR_{low;m,n} + MR_{high;m,n} = 1.$$

See Equation (6.11).

---

[1]The time index t used in Recommendation ITU-R M.1768 to characterize the market study parameters at different time intervals of the day is omitted here since the actual market data given in Report ITU-R M.2072 does not provide time-dependent values for the market study parameters.

4.3.2   Traffic distribution

The fourth step of the methodology flow chart in Figure 4.2 is to distribute the traffic of each
service category in each service environment and forecast year to different RATGs and radio
environments. This is done by computing the traffic distribution ratios that account for the
fraction of traffic that goes to each RATG and each radio environment. The traffic distribution
is performed separately for unicast and multicast traffic as well as for uplink and downlink
directions.

Unicast traffic

The traffic distribution ratio for unicast traffic of service category n in service environment m
to RATG rat and radio environment p is denoted by $\xi_{m,n,rat,p}$. If all traffic can be distributed,
the sum of the traffic distribution ratios for unicast traffic over the available RATGs and radio
environments for a given service category in a given service environment is equal to one:

$$\sum_{rat}\sum_{p} \xi_{m,n,rat,p} = 1. \tag{4.4}$$

The calculation of the traffic distribution ratios for unicast traffic involves the following
three stages:

(1) determination of possible combinations of service categories, service environments,
    radio environments and RATGs;

(2) calculation of intermediate traffic distribution ratios; and

(3) calculation of final traffic distribution ratios.

The input parameters to the first stage of the calculation of the traffic distribution ratios
for unicast traffic include mean service bit rate defined in Section 4.2.1, population coverage
percentage defined in Section 4.2.3, and application data rate and maximum supported
velocity defined in Section 4.2.4. Example values for the input parameters are given in
Section 6.4, Section 7.3.2 and Section 7.3.1, respectively. Outputs from the first stage include
possible combinations of service category, service environment, radio environment and
RATG in different forecast years which are denoted with marks '1' (possible) and '0'
(impossible). These are collected in the intermediate calculations performed in the MS Excel
tool of the spectrum requirement calculation methodology (see Section 5.4).

In the first stage, the availability of different radio environments and RATGs in different
service environments and forecast years is first determined for the different service categories.
This is done by comparing the data rate requirement of the service category with what the
different radio environments in different RATGs can offer. In essence, the mean service bit
rate of the service category is compared with the application data rate of the RATG in the
given radio environment. Moreover, the fact that not all radio environments are available in
all RATGs or all service environments is taken into account. This information is obtained
from the population coverage percentages of the different radio environments in different
service environments and the radio-related parameters of the RATGs. As a result from the
first stage, the traffic distribution algorithm presents the possible combinations of service

category, service environment, radio environment and RATG combinations in different years with mark '1' while the impossible combinations are denoted with '0'.

The second stage is to calculate the intermediate traffic distribution ratios for distributing the traffic to the available radio environments, determined in the first stage, based on the population coverage percentages of the different radio environments in the service environment defined in Section 4.2.3 and the mobility ratios of the service category in the given service environment defined in Section 4.2.1. The traffic in each service category is associated with corresponding mobility classes. The methodology defines the mobility classes: stationary/pedestrian, low and high mobility. Only certain radio environments can support certain mobility classes. The following mapping of mobility classes to radio environments is used:

- high mobility: macro cell only;

- low mobility: micro and macro cells;

- stationary/pedestrian: all radio environments.

The intermediate traffic distribution to radio environments follows the principle to use the radio environment with the lowest mobility support that satisfies the requirements determined in the first stage. This would mean that basically all stationary/pedestrian traffic would go to hot spot and pico cells, all low mobility to micro cells and all high mobility to macro cells. However, in practice each radio environment covers only a certain percentage of the total area of a particular service environment, which is defined as the population coverage percentage of the given radio environment in the given service environment. Therefore, the population coverage percentage puts a limit on the fraction of traffic that can be distributed to a given radio environment.

Let us denote the population coverage percentage of the hot spot, pico, micro and macro cell radio environments by $X_{hs;m}$, $X_{pico;m}$, $X_{micro;m}$ and $X_{macro;m}$, respectively, in service environment m. Note that $X_{macro;m} \equiv 1$. The mobility ratios of the offered traffic in the stationary/pedestrian and low mobility classes are denoted by $MR_{stat/ped;m,n}$ and $MR_{low;m,n}$, respectively, for service category n in service environment m as shown in Section 4.3.1. Using these parameters, the following algorithm is used to calculate the intermediate distribution ratios for the radio environments:

$$\xi_{pico\&hs;m,n} = \min\{X_{pico;m} + X_{hs;m}, MR_{stat/ped;m,n}\}, \tag{4.5}$$

$$\xi_{micro;m,n} = \min\{X_{micro;m}, MR_{low;m,n} + (MR_{stat/ped;m,n} - \xi_{pico\&hs;m})\}, \tag{4.6}$$

$$\xi_{macro;m,n} = 1 - \xi_{pico\&hs;m,n} - \xi_{micro;m,n}, \tag{4.7}$$

where $\min\{\cdots\}$ denotes the minimum operation. The traffic is further distributed between hot spots and pico cells in proportion to their population coverage percentages:

$$\xi_{hs;m,n} = \xi_{pico\&hs;m,n} \cdot \frac{X_{hs;m}}{X_{pico;m} + X_{hs;m}}, \tag{4.8}$$

$$\xi_{pico;m,n} = \xi_{pico\&hs;m,n} \cdot \frac{X_{pico;m}}{X_{pico;m} + X_{hs;m}}. \tag{4.9}$$

It is then obvious that

$$\sum_p \xi_{p;m,n} = 1.$$

In the third stage the final traffic distribution ratios are calculated based on the intermediate traffic distribution ratios from Equations (4.5)–(4.9) and the predetermined input parameter ξ_{rat} which is defined in Section 4.2.4 and listed in Figure 4.3. Example values for this input parameter are given in Section 7.3.2. The final traffic distribution ratios for unicast traffic are calculated from

$$\xi_{m,n,rat,p} = \xi_{p;m,n} \cdot \xi_{rat} \tag{4.10}$$

for $p \in$ {macrocell, micro cell, pico cell, hot spot}, where ξ_{rat} is the traffic distribution ratio for RATG rat among the available RATGs. Since

$$\sum_{rat} \xi_{rat} \leq 1,$$

we have

$$\sum_{rat} \xi_{m,n,rat,p} \leq \xi_{p;m,n},$$

where the sum is taken over the available RATGs. The final distribution ratios for unicast traffic fulfill Equation (4.4) if all traffic can be distributed. If part of the traffic is not distributed, the sum of the final traffic distribution ratios over the RATGs and radio environments is less than one.

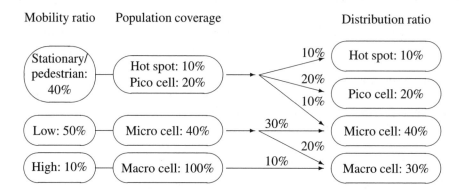

Figure 4.4 Example calculation of the traffic distribution ratios.

Figure 4.4 illustrates the calculation of the traffic distribution ratio for each radio environment using Equations (4.5)–(4.9). In this example, the mobility ratios are:

$$\text{MR}_{\text{stat/ped};m,n} = 0.4, \quad \text{MR}_{\text{low};m,n} = 0.5, \quad \text{MR}_{\text{high};m,n} = 0.1.$$

The population coverage for each radio environment is:

$$X_{\text{hs};m} = 0.1, \quad X_{\text{pico};m} = 0.2, \quad X_{\text{micro};m} = 0.4, \quad X_{\text{macro};m} = 1.$$

The traffic distribution proceeds as follows. First, the 40% traffic of stationary/pedestrian mobility class is destined to a hot spot/pico cell. However, since the population coverage of the hot spot and pico cell is 30%, only that amount is distributed to the hot spot/pico cell according to Equation (4.5), and the remaining traffic (10%) is distributed to a micro cell. The traffic in the hot spot/pico cell (30%) is further divided into the hot spot (10%) and the pico cell (20%) according to Equations (4.8) and (4.9). Then, the sum of the 50% traffic of low mobility class and the 10% traffic overflowed from the hot spot/pico cell is larger than the population coverage (40%) of a micro cell. Therefore, only 40% traffic is distributed to the micro cell according to Equation (4.6), and the remaining traffic (20%) is distributed to a macro cell. Finally, the sum of the 10% traffic of high mobility class and the 20% traffic overflowed from the micro cell is distributed to the macro cell according to Equation (4.7). Furthermore, if the available RATGs are only RATG 2 and RATG 3 with distribution ratio

$$\xi_{\text{RATG 2}} = 0.3, \quad \xi_{\text{RATG 3}} = 0.7$$

we have from Equation (4.10) that

$$
\begin{aligned}
\xi_{m,n,\text{RATG 2, hot spot}} &= 0.03, & \xi_{m,n,\text{RATG 3, hot spot}} &= 0.07, \\
\xi_{m,n,\text{RATG 2, pico cell}} &= 0.06, & \xi_{m,n,\text{RATG 3, pico cell}} &= 0.14, \\
\xi_{m,n,\text{RATG 2, micro cell}} &= 0.12, & \xi_{m,n,\text{RATG 3, micro cell}} &= 0.28, \\
\xi_{m,n,\text{RATG 2, macro cell}} &= 0.09, & \xi_{m,n,\text{RATG 3, macro cell}} &= 0.21,
\end{aligned}
$$

which sum to one.

Multicast traffic

The calculation of traffic distribution ratios for multicast traffic involves only two stages as follows:

(1) determination of possible combinations of service categories, service environments, radio environments and RATGs;

(2) calculation of final traffic distribution ratios.

The input parameters to the first stage of the calculation of the traffic distribution ratios for multicast traffic include the same parameters as for unicast traffic and additionally the parameter 'support for multicast' defined in Section 4.2.4. In the first stage, the service categories including multicast traffic are first identified from the market study which is introduced in Chapter 6. Then the availability of different radio environments and RATGs in different service environments and forecast years for the different service categories with multicast traffic is determined similarly as in the case of unicast traffic. However, at this point the availability of RATGs takes into account whether the given RATG can support multicast traffic. This is determined by the radio-related parameter 'support for multicast' of the different RATGs. Example values for this parameter are given in Section 7.3.1. The resulting possible combinations of service category, service environment, radio environment, and RATG in different forecast years are marked with '1' while the impossible combinations are denoted with '0'. These are collected in the intermediate calculations performed in the MS Excel tool of the spectrum requirement calculation methodology (see Section 5.4).

In the second stage, the final traffic distribution ratios for multicast traffic are determined by choosing the largest available cell for each service category with multicast traffic for each RATG in each service environment and forecast year based on the possible combinations derived in the first stage. The final distribution ratio is set equal to '1' to the largest cell of each RATG fulfilling the requirements while the other available combinations are turned to '0'. The resulting final distribution ratios for multicast traffic are denoted by $\xi_{m,n,rat,p}^{\text{Multicast}}$. The final traffic distribution ratios are calculated for each service category n with multicast traffic in each service environment m and available RATG rat in radio environment p according to

$$\xi_{m,n,rat,p}^{\text{Multicast}} = \begin{cases} 1 & \text{for } p \text{ with largest cell size} \\ 0 & \text{for other } p. \end{cases} \tag{4.11}$$

Thus, the multicast traffic is not divided among the different RATGs but all available RATGs deliver the whole multicast traffic in their largest available cell. Therefore, the multicast traffic does not take into account the input parameter 'distribution ratio among available RATGs'. The condition for the summation of final distribution ratios over the RATGs and radio environments presented in Equation (4.4) for unicast traffic does not hold for multicast traffic. Instead we have

$$\sum_p \xi_{m,n,rat,p}^{\text{Multicast}} = 1 \tag{4.12}$$

for those RATGs rat that can support the given multicast traffic.

4.3.3   Calculation of offered traffic

Using the traffic distribution ratios obtained in the preceding subsection, the traffic volume is distributed to different RATGs and different radio environments for each service category and each service environment. We then proceed to calculate the offered traffic per cell for each service category, each teledensity, each RATG, and each radio environment. At this point, the traffic is accumulated over the service environments which belong to the same teledensity. This process is also included in the fourth step of the methodology flow chart in Figure 4.2.

Session arrival rate per cell

Let us first calculate the session arrival rate per cell, which is done separately for unicast and multicast traffic. For unicast traffic, the session arrival rate per cell (sessions/s/cell) of service category n in service environment m is distributed to RATG rat and radio environment p according to the formula

$$P_{m,n,rat,p} = \xi_{m,n,rat,p} \cdot U_{m,n} \cdot Q_{m,n} \cdot A_{d,p}, \tag{4.13}$$

where $U_{m,n}$ is the user density (users/km²) and $Q_{m,n}$ is the session arrival rate per user (sessions/s/user) of service category n in service environment m explained in Section 4.3.1. $A_{d,p}$ is the cell area (km²/cell) of radio environment p in teledensity d, where d is uniquely determined by m. Finally, $\xi_{m,n,rat,p}$ is the traffic distribution ratio (dimensionless) of service category n in service environment m to RATG rat in radio environment p.

Multicast traffic is transmitted to multiple receivers simultaneously using a shared radio resource. Therefore, the user population is assumed assumes to have a negligible effect. Hence we may assume there to be a single user in a cell, i.e.

$$U_{m,n} \cdot A_{d,p} = 1 \quad \text{for multicast traffic of service category } n. \tag{4.14}$$

As a result, for multicast traffic, the session arrival rate per cell (sessions/s/cell) of service category n in service environment m is distributed to RATG rat and radio environment p according to the formula:

$$P_{m,n,rat,p} = \xi_{m,n,rat,p}^{\text{Multicast}} \cdot Q_{m,n}. \tag{4.15}$$

The conversational and streaming traffic classes (SC 1–10) are assumed to be served with circuit switching while the interactive traffic and background classes (SC 11–20) are served with packet switching. These two schemes require different parameters to characterize the demand. The traffic values are therefore calculated separately for circuit-switched and packet-switched service categories.

Offered traffic and mean service bit rate for circuit-switched service categories

For circuit-switched service categories (SC 1–10), the traffic distribution process yields the session arrival rate per cell $P_{m,n,rat,p}$ as explained above. Also the average session duration $\mu_{m,n}$ (s/session) is available from market study as explained in Section 4.3.1. Then the product $P_{m,n,rat,p} \cdot \mu_{m,n}$ represents the offered traffic per cell (Erlang/cell) of service category n in service environment m for RATG rat in radio environment p. The unit 'Erlang' for the offered traffic is physically dimensionless. It represents the fraction of time that the communication channel is occupied. The aggregate offered traffic per cell of service category n in teledensity d for RATG rat in radio environment p is then calculated from

$$\rho_{d,n,rat,p} = \sum_{m \in d} P_{m,n,rat,p} \cdot \mu_{m,n}, \tag{4.16}$$

where the summation is taken over the service environments m which belong to teledensity d.

The mean service bit rate $r_{m,n}$ (bit/s) of service category n in service environment m is available from market study as explained in Section 4.3.1. Then the product $P_{m,n,rat,p} \cdot \mu_{m,n} \cdot r_{m,n}$ represents the average number of bits transmitted per second per cell (bit/s/cell). The average values for the mean service bit rate (bit/s) of service category n in teledensity d for RATG rat in radio environment p are then calculated from

$$r_{d,n,rat,p} = \frac{\sum_{m \in d} P_{m,n,rat,p} \cdot \mu_{m,n} \cdot r_{m,n}}{\rho_{d,n,rat,p}}. \tag{4.17}$$

Offered traffic for packet-switched service categories

For packet-switched service categories (SC 11–20), we interpret $P_{m,n,rat,p} \cdot \mu_{m,n} \cdot r_{m,n}$ as the average number of bits that arrive per second in a cell (bit/s/cell) of service category n in service environment m for RATG rat in radio environment p. Therefore, the aggregate offered traffic per cell (bit/s/cell) of service category n in teledensity d for RATG rat in radio environment p is calculated from

$$T_{d,n,rat,p} = \sum_{m \in d} P_{m,n,rat,p} \cdot \mu_{m,n} \cdot r_{m,n}, \tag{4.18}$$

where the summation is taken over the service environments m which belong to teledensity d.

The output from traffic calculation in the fourth step of the methodology flow chart is the offered traffic in Equation (4.16) and the mean service bit rate in Equation (4.17) for circuit-switched service categories, and the offered traffic in Equation (4.18) for packet-switched service categories. These are calculated separately for RATG 1 and RATG 2 for unicast and multicast traffic in each teledensity, each radio environment and each forecast year for uplink and downlink directions.

4.3.4   Required capacity for circuit-switched service categories

In the fifth step of the methodology flow chart in Figure 4.2, the required system capacity needed to serve the offered traffic per cell while fulfilling the QoS requirements of each service category n is determined for each RATG rat and radio environment p in each teledensity d and in uplink/downlink directions for the forecast year. The required system capacity per cell, measured in bit/s/cell, is calculated separately for circuit-switched and packet-switched service categories. The number of circuit-switched service categories is denoted by N^{cs}, while the number of packet-switched service categories is denoted by N^{ps}.

The results of these calculations are the required system capacity $C^{cs}_{d,rat,p}$ for circuit-switched categories and $C^{ps}_{d,rat,p}$ for packet-switched categories, both measured in bit/s/cell. Namely, $C^{cs}_{d,rat,p}$ represents the system capacity that is required to fulfill the QoS requirements of all circuit-switched service categories in teledensity d, RATG rat and radio environment p. Also, $C^{ps}_{d,rat,p}$ is the system capacity that is required to fulfill the QoS requirements of all packet-switched service categories in teledensity d, RATG rat and radio environment p. The indices for the link direction and the forecast year are implicit here. The sum of $C^{cs}_{d,rat,p}$ and $C^{ps}_{d,rat,p}$ is the total capacity requirement, denoted by $C_{d,rat,p}$, for all service categories in teledensity d, RATG rat and radio environment p, as shown in Equation (4.38) below.

In this section, we show the method for finding the required system capacity $C^{cs}_{d,rat,p}$ for circuit-switched service categories (the method for packet-switched service categories is given in Section 4.3.5). This is determined in terms of the number of service channels needed to achieve the specified values of blocking probability for all circuit-switched service categories and the resultant system capacity for each teledensity, RATG and radio environment.

The input parameters for this calculation are as follows:

- offered traffic per cell $\rho_{d,n,rat,p}$ (Erlang/cell) of service category n in teledensity d for RATG rat in radio environment p, given in Equation (4.16);

- mean service bit rate $r_{d,n,rat,p}$ (bit/s) of service category n in teledensity d for RATG rat in radio environment p, given in Equation (4.17);

- maximum allowable blocking probability π_n (dimensionless) for service category n, described in Section 4.2.1.

In the following, we denote $\rho_{d,n,rat,p}$ and $r_{d,n,rat,p}$ by ρ_n and r_n, respectively, to improve the readability. Taking trunking gain into account, the well-known Erlang-B formula for a loss system with a single class of calls can be extended to the case of multiple traffic classes which also allows simultaneous occupation of several channels by each call (multidimensional Erlang-B formula with multiple server occupation). We assume that calls of N^{cs} classes share the set of s channels and that each call of class n requires m_n channels simultaneously, $1 \leq n \leq N^{cs}$. If an arriving call of class n finds less than m_n idle channels, then it is blocked and lost; let $\boldsymbol{m} \equiv (m_1, m_2, \ldots, m_{N^{cs}})$. Calls of class n arrive in a Poisson process independent of other classes, and they have exponentially distributed holding times such that the offered traffic of class n is ρ_n. All channels used by a call are released at the end of the holding time. Let the system state be $\boldsymbol{k} \equiv (k_1, k_2, \ldots, k_{N^{cs}})$, where k_n is the number of calls of class n currently using channels. According to Equation (A.42) in Appendix A.3, the steady-state probability mass function has a simple product form:

$$P(\boldsymbol{k}) = \frac{1}{G(s)} \prod_{n=1}^{N^{cs}} \frac{(\rho_n)^{k_n}}{k_n!} \tag{4.19}$$

with

$$G(j) = \sum_{\{k:0 \leq m \cdot k \leq j\}} \prod_{n=1}^{N^{cs}} \frac{(\rho_n)^{k_n}}{k_n!}, \quad 1 \leq j \leq s \tag{4.20}$$

where

$$\boldsymbol{m} \cdot \boldsymbol{k} \equiv \sum_{n=1}^{N^{cs}} m_n k_n$$

is the number of channels being used when the system state is $\boldsymbol{k}$. The blocking probability for calls of class n is then given by

$$P_n^B(s) = \sum_{\{k:m \cdot k > s - m_n\}} P(\boldsymbol{k}) = 1 - \frac{G(s - m_n)}{G(s)}. \tag{4.21}$$

Since a brute force computation of $G(j)$ by Equation (4.20) involves computational difficulties, several efficient algorithms have been developed. Among them a recursive algorithm, so-called *Kaufman–Roberts algorithm* (Kaufman 1981; Roberts 1981), is simple and computationally preferable. Their algorithm is further modified in order to be suitable for repetitive calculation in the inverse problem of determining the system capacity in order to satisfy the user's requirement on the blocking probabilities (Takagi et al. 2006). Namely, starting with $G(0) = 1$, we calculate $G(j)$, $j = 1, 2, \ldots, s$, recursively by

$$G(j) = \frac{1}{j} \left[\sum_{l=0}^{j-1} G(l) + \sum_{n=1}^{N^{cs}} m_n \rho_n G(j - m_n) \right], \tag{4.22}$$

where $G(l) = 0$ for $l < 0$. This algorithm yields the blocking probabilities for systems with up to s channels all at once with $O(N^{cs}s)$ computational time and $O(s)$ memory requirement. Equations (4.19)–(4.22) are derived in Appendix A.3.

The above model and algorithm are used to compute the blocking probability for each of N^{cs} service categories when the total number of channels, s, is given. By the inverse method, the total number of channels is calculated so as to meet the condition on the blocking probability for every category required by the user. The system capacity is obtained by multiplying the required total number of channels by the bit rate per channel.

Let r (bit/s) be the unit of service bit rate per channel. When the service bit rate for category n is r_n, the parameter m_n to be used in the above formula is given by:

$$m_n = \lceil r_n/r \rceil, \quad 1 \le n \le N^{\text{cs}}, \tag{4.23}$$

where $\lceil x \rceil$ denotes the least integer greater than or equal to x (ceiling function). This means that the number of channels is counted using r as the unit data rate for each service category.

Let π_n be the blocking probability of service category n requested by the user. Then the required number of channels per cell, $s_{\min}$, is obtained as the smallest s that satisfies the requirement

$$P_n^{\text{B}}(s) < \pi_n \tag{4.24}$$

for all $1 \le n \le N^{\text{cs}}$ simultaneously. Finally, the required system capacity $C^{\text{cs}}_{d,rat,p}$ (bit/s/cell) for all the circuit-switched service categories is given by

$$C^{\text{cs}}_{d,rat,p} = s_{\min} \cdot r. \tag{4.25}$$

4.3.5 Required capacity for packet-switched service categories

In this section, we show the method for finding the required system capacity $C^{\text{ps}}_{d,rat,p}$ for packet-switched service categories. This method was published in a first version by Irnich and Walke (2004). The complete method was published by Irnich *et al.* (2005). It was also applied in (Irnich and Walke 2005) to calculate the spectrum requirements of IMT-2000 systems, using the delay criteria specified in ITU-R Recommendation M.1079. It is shown there that the method provides similar results as the methodology described in Chapter 3, but on a much more solid ground.

Let us denote by N^{ps} the number of packet-switched service categories. The input parameters for this calculation are as follows:

- offered traffic per cell $T_{d,n,rat,p}$ (bit/s/cell) of service category n in teledensity d for RATG rat in radio environment p, given in Equation (4.18);

- mean packet size s_n (bit/packet) of service category n, described in Section 4.2.1;

- second moment of packet size $s_n^{(2)}$ (bit²/packet²) of service category n, described in Section 4.2.1;

- maximum allowable mean packet delay Δ_n (s/packet) for service category n, described in Section 4.2.1.

The arrival rate of IP packets refers to the average number of IP packets that are transmitted per unit time per cell (packets/s/cell). Thus the packet arrival rate per cell of service category n in teledensity d for RATG rat in radio environment p is obtained by dividing the offered traffic by the mean packet size:

$$\lambda_{d,n,rat,p} = \frac{T_{d,n,rat,p}}{s_n}. \tag{4.26}$$

In order to improve the readability, the indices d, rat and p are omitted so that $\lambda_{d,n,rat,p}$ is simply denoted by λ_n until the end of this section ($1 \le n \le N^{\text{ps}}$).

Let C (bit/s) be the capacity of the channel, i.e. the number of bits transmitted per unit time. Then the mean b_n (s/packet) and the second moment $b_n^{(2)}$ (s^2/packet^2) of the transmission time (service time) for an IP packet of service category n are given by

$$b_n = \frac{s_n}{C}; \quad b_n^{(2)} = \frac{s_n^{(2)}}{C^2}. \tag{4.27}$$

Unlike the calls of circuit-switched service categories which are blocked and lost if all channels are busy when they arrive, the IP packets of packet-switched service categories are queued when the channel is used at the time of arrival. Therefore we want a queueing model with multiple classes of customers (packets) with class-dependent generally distributed service times for which a simple formula of the mean delay is available. One such model is a classical M/G/1 nonpreemptive priority queue. In this model, a single server (transmission channel) attends all classes of customers. It is assumed that there is a linear order of priority among the classes. When the server becomes available, it selects for the next service one of the customers whose priority is the highest among those waiting in the queue. Customers of the same class are served on the first-come first-served (FCFS) basis. The nonpreemptive service means that once the service to a customer is started it is not disrupted until completion even if any customers of higher priority arrive during the service. A simple formula, due to Cobham (1954), is available for the mean delay, i.e. the time from arrival to service completion, of each class. For the convenience of the reader, the derivation of the mean waiting time in the M/G/1 nonpreemptive priority queue is shown in Appendix A.4.

Although this priority operation may not model exactly the practical scheduling of IP packet transmission in real wireless systems, we adopt the M/G/1 nonpreemptive priority queue to determine the required system capacity for packet-switched service categories. The main reason for this choice is the availability of Cobham's formula. Hence we assume that service categories of IP packets correspond to priority classes in the queueing model. There are N^{ps} classes indexed as $n = 1, 2, \ldots, N^{\text{ps}}$. Packets of class n arrive in a Poisson process with rate λ_n. Its service time may have general distribution with mean b_n and the second moment $b_n^{(2)}$ as given above. Class n has priority over class i if and only if $n < i$; thus class 1 is the highest priority and class N^{ps} the lowest.

In this setting, from Cobham's formula, given in Equation (A.89) in Appendix A.4, we have the mean delay of an IP packet of service category n as

$$D_n(C) = \frac{\sum_{i=1}^{N^{\text{ps}}} \lambda_i b_i^{(2)}}{2(1 - \sum_{i=1}^{n-1} \lambda_i b_i)(1 - \sum_{i=1}^{n} \lambda_i b_i)} + b_n$$

$$= \frac{\sum_{i=1}^{N^{\text{ps}}} \lambda_i s_i^{(2)}}{2(C - \sum_{i=1}^{n-1} \lambda_i s_i)(C - \sum_{i=1}^{n} \lambda_i s_i)} + \frac{s_n}{C}. \tag{4.28}$$

The required system capacity C_n is obtained as the smallest C that satisfies the requirement

$$D_n(C) < \Delta_n. \tag{4.29}$$

Note that Equation (4.28) can be written as

$$\left(D_n(C) - \frac{s_n}{C}\right)\left(C - \sum_{i=1}^{n-1} \lambda_i s_i\right)\left(C - \sum_{i=1}^{n} \lambda_i s_i\right) = \frac{1}{2}\sum_{i=1}^{N^{\text{ps}}} \lambda_i s_i^{(2)}. \tag{4.30}$$

Since $D_n(C)$ is a monotonously decreasing function in C, it follows that C_n is given as a solution to the cubic equation

$$f_n(x) = 0, \qquad (4.31)$$

where

$$f_n(x) := 2(x\Delta_n - s_n)\left(x - \sum_{i=1}^{n-1}\lambda_i s_i\right)\left(x - \sum_{i=1}^{n}\lambda_i s_i\right) - x\sum_{i=1}^{N^{ps}}\lambda_i s_i^{(2)}$$

$$= \alpha_n x^3 + \beta_n x^2 + \gamma_n x + \delta_n \qquad (4.32)$$

with the coefficients α_n, β_n, γ_n, and δ_n given by

$$\alpha_n = 2\Delta_n,$$

$$\beta_n = -2\left[\Delta_n\left(\sum_{i=1}^{n-1}\lambda_i s_i + \sum_{i=1}^{n}\lambda_i s_i\right) + s_n\right],$$

$$\gamma_n = 2\left[\Delta_n\left(\sum_{i=1}^{n-1}\lambda_i s_i\right)\left(\sum_{i=1}^{n}\lambda_i s_i\right) + s_n\left(\sum_{i=1}^{n-1}\lambda_i s_i + \sum_{i=1}^{n}\lambda_i s_i\right)\right] - \sum_{i=1}^{N^{ps}}\lambda_i s_i^{(2)},$$

$$\delta_n = -2s_n\left(\sum_{i=1}^{n-1}\lambda_i s_i\right)\left(\sum_{i=1}^{n}\lambda_i s_i\right). \qquad (4.33)$$

Note that

$$f_n\left(\sum_{i=1}^{n}\lambda_i s_i\right) < 0; \quad f_n(+\infty) = +\infty. \qquad (4.34)$$

Therefore, Equation (4.31) always has a root $x = C_n$, a real number, such that

$$\sum_{i=1}^{n}\lambda_i s_i < C_n. \qquad (4.35)$$

In fact, this is the stability condition that the queue of class n packets does not grow indefinitely.

The solution to the cubic equation in (4.31) can be found either numerically, e.g. by Newton's method, or algebraically, e.g. by *Cardano's formula*. In the MS Excel tool (Chapter 5), the solution by Cardano's formula is implemented.

After the channel capacity C_n that satisfies the requirement (4.29) for service category n has been obtained for all categories ($1 \leq n \leq N^{ps}$), the required system capacity $C_{d,rat,p}^{ps}$ (bit/s/cell) for all the packet-switched service categories is given by

$$C_{d,rat,p}^{ps} = \max\{C_1, C_2, \dots, C_{N^{ps}}\}. \qquad (4.36)$$

4.3.6 Spectrum results

Recall that the calculation in the fifth step of the methodology flow chart in Figure 4.2, shown in Sections 4.3.4 and 4.3.5, is done separately for:

- each teledensity d;

- each RATG rat;

- each radio environment p;

- uplink and downlink directions;

- forecast years.

In the notation of required capacities $C_{d,rat,p}^{cs}$ and $C_{d,rat,p}^{ps}$, the indices for link direction and forecast year are implicit.

In the sixth step of the methodology flow chart in Figure 4.2, the capacity requirements of circuit-switched and packet-switched service categories for unicast and multicast traffic in RATG 1 and RATG 2 are processed to get the corresponding spectrum requirements in different teledensities, RATGs, radio environments and forecast years.

To do so, the capacity requirements of the uplink and downlink directions are first summed up. Let us redefine by $C_{d,rat,p}^{cs}$ and $C_{d,rat,p}^{ps}$ the resulting capacity requirements for packet-switched and circuit-switched service categories, respectively, as follows:

$$C_{d,rat,p}^{cs} = C_{d,rat,p,\text{uplink}}^{cs} + C_{d,rat,p,\text{downlink}}^{cs}$$

$$C_{d,rat,p}^{ps} = C_{d,rat,p,\text{uplink}}^{ps} + C_{d,rat,p,\text{downlink}}^{ps} \qquad (4.37)$$

where the index for the forecast year is still implicit. These are added together to obtain the total capacity requirement $C_{d,rat,p}$ (bit/s/cell) of all service categories in teledensity d for RATG rat in radio environment p according to

$$C_{d,rat,p} = C_{d,rat,p}^{cs} + C_{d,rat,p}^{ps}. \qquad (4.38)$$

The capacity requirements are now converted to the spectrum requirements simply by dividing the former by the corresponding spectral efficiency values. Specifically, the spectrum requirement $F_{d,rat,p}$ (Hz) of RATG rat in teledensity d and radio environment p is calculated from

$$F_{d,rat,p} = C_{d,rat,p}/\eta_{d,rat,p} \qquad (4.39)$$

where $\eta_{d,rat,p}$ is the area *spectral efficiency* (bit/s/Hz/cell) of RATG rat in teledensity d and radio environment p. In the case of mobile multicast capacity requirements, the corresponding spectrum requirement $F_{d,rat,p}^{\text{Multicast}}$ (Hz) is calculated separately, using the appropriate spectral efficiency $\eta_{d,rat,p}$ values. This spectrum requirement is added to the unicast spectrum requirements to obtain

$$F_{d,rat,p} \leftarrow F_{d,rat,p} + F_{d,rat,p}^{\text{Multicast}}. \qquad (4.40)$$

The seventh step of the methodology flow chart in Figure 4.2 is to apply necessary adjustments to the spectrum requirements. They are adjusted with the number of overlapping network deployments and the minimum spectrum deployment per operator per radio environment. First, the unadjusted spectrum requirement per operator is calculated from

$$F_{d,rat,p} \leftarrow F_{d,rat,p}/N_{o}, \qquad (4.41)$$

where N_{o} is the number of overlapping network deployments (dimensionless).

Spectrum can only be used with the granularity of the minimum spectrum deployment per operator and radio environment which is denoted as $\text{MinSpec}_{rat,p}$ (Hz). The spectrum requirement is adjusted with the minimum spectrum deployment according to

$$F_{d,rat,p} = \text{MinSpec}_{rat,p}\lceil F_{d,rat,p}/\text{MinSpec}_{rat,p}\rceil, \tag{4.42}$$

where $\lceil x \rceil$ denotes the least integer greater than or equal to x (ceiling function).

In the eighth step of the methodology flow chart of Figure 4.2, the spectrum requirements are aggregated over radio environments and the guard bands between operators are taken into account. In combining the spectrum requirements over radio environments, the pico cells and hot spots are assumed to be spatially non-coexisting, while the macro cells and micro cells are assumed to be spatially coexisting with the pico cells and hot spots. Therefore, the spectrum requirement $F_{d,rat}$ (Hz) in teledensity d for RATG rat is calculated from

$$F_{d,rat} = F_{d,rat,\text{macro}} + F_{d,rat,\text{micro}} + \max\{F_{d,rat,\text{pico}}, F_{d,rat,\text{hs}}\}, \tag{4.43}$$

where $\max\{\cdots\}$ denotes the maximum operation. Then, the total required spectrum for all operators including the guard bands between operators becomes

$$F_{d,rat} \leftarrow N_{\text{o}} \cdot F_{d,rat} + (N_{\text{o}} - 1) \cdot F_{rat}^{\text{G}} \tag{4.44}$$

where F_{rat}^{G} (Hz) is the guard band between operators for RATG rat.

In the final step (Step 9) of the methodology flow chart in Figure 4.2, the total spectrum requirements of RATG 1 and RATG 2 are taken as the maximum over the teledensities $d \in$ {dense urban, suburban, rural}.[2] Thus we obtain the spectrum requirement F_{rat} (Hz) by

$$F_{rat} = \max_{d} F_{d,rat}. \tag{4.45}$$

The final spectrum requirements of RATG 1 and RATG 2 are given for the three different forecast years.

4.4 Summary of Methodology for IMT-Advanced

The spectrum calculation methodology presented in this chapter yields the spectrum requirements for RATG 1 (pre-IMT, IMT-2000, and future development of IMT-2000) and RATG 2 (IMT-Advanced) in the three forecast years 2010, 2015 and 2020. The methodology accommodates a complex mix of services from market studies using service categories with different traffic volumes and QoS constraints which characterize the predicted demand of future wireless services. The methodology is technology neutral and handles emerging as well as established systems using a RATG approach with a limited set of radio parameters. The four RATGs considered cover all relevant radio access technologies that are assumed to carry the future traffic predicted in the market studies.

The methodology processes the market study traffic volumes to obtain predictions of the traffic demand, data rates and user mobilities. The methodology distributes the traffic from

---

[2]The final spectrum requirement calculation presented here does not consider the time-dependency of the spectrum requirements or the use of flexible spectrum use (FSU) presented in Recommendation ITU-R M.1768, because they are not used in the actual calculations performed in Report ITU-R M.2078.

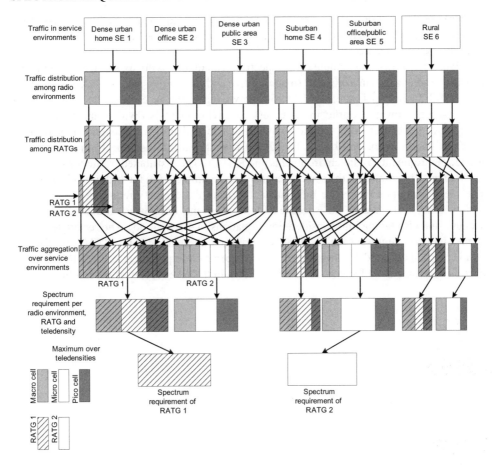

Figure 4.5 Traffic distribution and spectrum requirement calculation.

the market studies to four RATGs and four radio environments using technical and market-related information. The traffic distributed to RATG 1 and RATG 2 is further processed in the methodology by transforming the traffic volumes into capacity requirements, using separate calculation algorithms for packet-switched and circuit-switched service categories. The methodology defines two separate capacity calculation algorithms in order properly to model the services with different characteristics. Some service categories require guaranteed bit rates and they are treated in the network more as in 'circuit-switched' manner. Thus they are modeled using the Erlang-based formula. Other service categories have a more truly packet-oriented nature in the network. For those services a queueing model is applied.

The capacity calculation algorithms take into account the gain in multiplexing among circuit-switched service categories, as well as among packet-switched service categories with differing QoS characteristics. We have not considered multiplexing of circuit-switched and packet-switched service categories. The capacity requirements are processed with spectral efficiency values to obtain the initial spectrum requirements. The methodology further considers factors relevant for practical network deployments to adjust the spectrum requirements, and calculates the final spectrum requirements of RATG 1 and RATG 2 in the three different forecast years.

Figure 4.5 shows an example of the spectrum calculation with six service environments (SE 1–6), three radio environments (macro, micro and pico cells), and two RATGs (RATG 1 and RATG 2). In this figure, the first row shows the traffic in each service environment. In the second row the traffic in each service environment is divided among the radio environments. The traffic in each radio environment is further distributed among the RATGs, which is shown in the third row. The fourth row shows the resulting traffic in different service environments, radio environments and RATGs. The distributed traffic for the service environments belonging to the same teledensity is accumulated in the fifth row of the figure. The spectrum requirements are then calculated from the capacity requirements according to Equation (4.38), taking the spectral efficiency of the RATGs in Equation (4.39) into account, and also taking into consideration Equations (4.40) to (4.45). The rectangles shown in the sixth row represent the resulting spectrum requirements for the RATGs in different radio environments and teledensities. The final spectrum requirement of each RATG is the maximum among the teledensities which is shown in the seventh row of the figure.

5

Calculation Tool Package

Marja Matinmikko, Jörg Huschke and Jussi Ojala

The spectrum requirement calculation methodology for the future development of IMT-2000 and IMT-Advanced systems presented in Chapter 4 has been implemented into a software tool. The tool is called 'SPEctrum requirement calCULATOR for the future development of IMT-2000 and IMT-Advanced' (SPECULATOR). The tool has been implemented in Microsoft (MS) Excel to allow wide usability and to achieve transparency in calculations. The tool is publicly available on the ITU web site 'http://www.itu.int/ITU-R/study-groups/docs/speculator.doc'. ITU used the tool to calculate the estimates of the spectrum requirements of IMT-Advanced at World Radiocommunication Conference 2007 (WRC-07).

A reader not interested in operating the tool might wish to skip this chapter.

5.1   Description and Use of Software Tool

The MS Excel tool for implementing the ITU spectrum calculation methodology for the future development of IMT-2000 and IMT-Advanced consists of 27 worksheets and seven modules of macros. The MS Excel worksheets include input parameter values, intermediate calculation results from spreadsheet calculations and macro calculations, and final spectrum requirements with some charts. Some of the calculations are performed on the worksheets while macros perform the more complicated calculations. Macros are also used to read input values from the worksheets and write output values to the worksheets.

The usage of the tool requires MS Excel version 2000 or later. The tool can be executed on computers that have 256 MB or more RAM. To enable the usage of the macros, the security level of MS Excel has to be set to 'Medium' or lower, otherwise the macros are disabled when the workbook is opened, which prevents the use of the tool.

The software tool includes all the input parameter values from Report ITU-R M.2078. Therefore, it can be directly used to perform the calculations. The tool includes the market data from Report ITU-R M.2072, and performs all the necessary pre-processing needed for the calculations. Many parameters in the tool have an inherent technical relationship with each other that needs to be kept in mind. Therefore, it is often not reasonable to change a single parameter value without adapting other related parameters.

The tool is implemented with a view to performing the spectrum requirement calculations as described in Recommendation ITU-R M.1768. This means that the inputs and outputs all follow the agreed format and the procedure. Therefore, the tool should not be modified. It is important to note that for most of the worksheets, adding or changing the order of columns or rows will make the tool unusable.

The tool can be divided into four parts: front sheet, input parameters, intermediate calculations and outputs. The following sections describe in more detail the structure of the tool and the four different parts.

5.2 Front Sheet of Software Tool

The software tool is executed from the front sheet, called 'Main', which is the core of the MS Excel tool. The front sheet is shown in Figure 5.1. The front sheet includes buttons for performing the calculations as well as the spectrum requirement results. The front sheet consists of six parts:

- Major parameters

- Calculate the spectrum requirement

- Output

- Warnings

- Time-shifted spectrum requirement

- Unadjusted spectrum requirement.

The Major parameters section includes two parameters which have a high effect on the spectrum requirements. These parameters are the number of overlapping network deployments and the cell area, denoted as 'number of network deployments' and 'sector area' in the MS Excel tool. The number of network deployments denotes the number of independent overlapping networks of the same RATG that do not share the spectrum. Separate values are given for RATG 1 and RATG 2. The sector area is the cell size of the radio environments in different teledensities.

The second section, i.e. 'calculate the spectrum requirement', includes seven buttons for performing the different calculation steps and one button for performing all calculation steps. After all input values have been put into the worksheets, it is possible to use the tool from the front sheet using the buttons. Each of the buttons executes different macros in the tool. Alternatively, the button 'Run All' can be used. The processing status is also shown. The execution of all calculations can take several minutes. When a button is pressed and the processing is started, the status on the right-hand side of the button becomes 'RUNNING'. After the processing is completed without errors, the processing status becomes 'OK'. In case of errors in execution, the status becomes 'ERROR'. The Methodology flow chart is

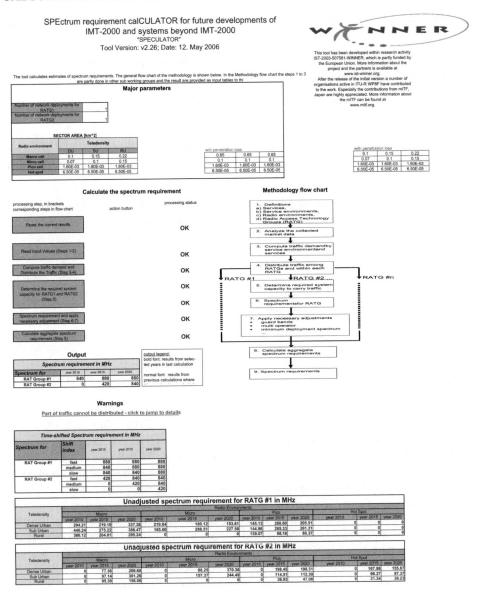

Figure 5.1 Front sheet of the tool for ITU spectrum calculation methodology.

also presented to show the connection between the processing steps of the tool and the steps in the methodology. This flow chart corresponds to the flow chart shown in Figure 4.2.

The individual buttons for performing the calculations include 'Reset', 'Read Input Values', 'Distribute Traffic', 'PS-Capacity', 'CS-Capacity', 'Apply Adjust.' and 'Spectrum Req.'. The macros associated with the buttons are collected in the module called

'MainModule' which contains the main command macros which call macros from other modules. The actions of the individual buttons include the following.

- The first button 'Reset' calls macros which reset the values in the worksheets to initialize the calculations.

- The second button 'Read Input Values' calls the macros in 'ReadInput' module which read the input parameter values from the market studies.

- The third button 'Distribute Traffic' calls the macros in 'DistributionModule' and 'OutputTrafficModule' which perform the traffic distribution and calculate the distributed traffic.

- The fourth button 'PS-Capacity' calls the macros in 'PSCapacityModule' which calculates the capacity requirements for packet-switched service categories.

- The fifth button 'CS-Capacity' calls the macros in 'CSCapacityModule' which calculates the capacity requirements for circuit-switched service categories.

- The sixth button 'Apply Adjust.' applies adjustments on the spectrum requirements on the worksheet without separate macros.

- The seventh button 'Spectrum Req.' calculates the final spectrum requirements and prints them on the front sheet 'Main'.

It is important that the buttons are executed in the correct order, i.e. in the order the buttons appear on the front sheet.

The third section, i.e. 'Output', shows a table of the final spectrum requirements of RATG 1 and RATG 2 in the three different forecast years 2010, 2015 and 2020. The calculations are usually performed for all three forecast years at the same time but the calculations can also be run separately for the different years. In that case, only the newly calculated results are updated to the table. The updated results are shown in bold font while the old results are shown in normal font.

The fourth section, i.e. 'Warnings', shows errors in the calculations. For example, if part of traffic cannot be distributed in Step 4 of the methodology flow chart, a warning message 'Part of traffic cannot be distributed – click to jump to details' is written. If the calculations are performed for only one or two of the three forecast years, the warning 'Spectrum requirements for deselected years were not updated' is written.

The fifth section, i.e. 'Time-shifted spectrum requirement', presents the final spectrum requirements of RATG 1 and RATG 2 in so-called *time-shifted format*. The time-shifted format defines three development scenarios for the deployment of IMT systems: fast, medium and slow deployment. The three scenarios are used to model the differences in the speed of market development and deployment of the new systems in different parts of the world.

The sixth section, i.e. 'Unadjusted spectrum requirement', gives the intermediate spectrum requirement results which show the unadjusted spectrum requirements of RATG 1 and RATG 2 in different teledensities and different radio environments in different forecast years. This corresponds to the output from Equation (4.40). The unadjusted spectrum requirements are shown on the front sheet to show which teledensities and radio environments dominate the spectrum requirements before any adjustments are made.

5.3   Inputs to Software Tool

The input parameters to the calculations including the market and service category parameters, radio-related parameters and other parameters are given in worksheets 2–13 of the MS Excel tool. In the following, the worksheets containing the input parameters are described. The parameter values included in the tool are from Report ITU-R M.2078.

Worksheet 2

The 'Market-Setting' worksheet (sheet 2) contains the year selector and the market attribute settings. The year selector is used to choose the forecast years for which the calculations are performed. The calculations can be run separately for 2010, 2015 and 2020 or combinations thereof. Market attribute settings choose the actual values for the market parameters within the ranges given in Report ITU-R M.2072. Since the market report is a collection of different market studies, only the ranges of value are given there instead of specific values. The user can select individual values within the ranges by defining percentages for each service category in each forecast year which determine the actual values within the ranges. The percentage '0%' represents the lower limit of the range and '100%' is the upper limit of the range. The percentage is available for the parameters: user density, session arrival rate per user, mean service bit rate and average session duration. The percentage approach is further illustrated in Section 6.4. The mobility ratios are selected differently using mobility scenarios. The tool offers three sets of mobility scenarios including lowest, middle and highest mobility scenarios. The mobility scenarios can be selected by the numbers 'one' to 'three', where 'one' corresponds to the lowest mobility scenario and 'three' to the highest mobility scenario.

Worksheet 3

In the 'RATG-DistRatio-Input' worksheet (sheet 3), the user can give values to the distribution ratios among available RATGs for three different forecast years. This parameter is used in the unicast traffic distribution if several RATGs are available for a given service category in a service environment and radio environment. The parameter defines the splitting of unicast traffic among these available RATGs. It should be noted that RATG 4 can only support multicast type of traffic which is not distributed among different RATGs but that all multicast traffic is carried by all supporting RATGs. Therefore, separate values are not given to RATG 4.

Worksheet 4

The 'SE-Input' worksheet (sheet 4) gives the mapping of radio environments and service environments, i.e. it shows the availability of different radio environments in the different service environments for three different forecast years. The worksheet also contains the values for the population coverage percentage for the three different forecast years.

Worksheet 5

The 'RATG1&RATG2Def-Input' worksheet (sheet 5) includes the radio parameters for RATG 1 and RATG 2. Specific values are needed for application data rate, maximum supported velocity, guard band between operators, minimum deployment per operator and radio environment, support for multicast, and number of overlapping network deployments. The number of overlapping network deployments is read from 'Main' sheet. Moreover, the application data rate is given separately for the three different forecast years in order to model the existence of the different RATGs in different forecast years. If a given RATG does not exist in a given year, the application data rate is set to zero in all radio environments for the corresponding year.

Worksheet 6

The 'RATGEff-Input' worksheet (sheet 6) shows tables which give the spectral efficiency values for RATG 1 and RATG 2 in different radio environments and teledensities. Separate tables are given for the three forecast years and for unicast and multicast traffic.

Worksheet 7

The 'RATG3&4Def-Input' worksheet (sheet 7) includes the radio parameters for RATG 3 and RATG 4. Specific values are needed for the two RATGs for application data rate, support for multicast and maximum supported velocity.

Worksheet 8

The 'RATGBearer-Input' worksheet (sheet 8) shows which of the service categories are served in circuit-switched or packet-switched scheme. The cells on this worksheet are locked so that the user cannot change the values because the methodology defines fixed mapping of service categories to the transmission scheme, that is to say SC 1–10 are circuit-switched and SC 11–20 are packet-switched. Therefore, the values in this table are for information only.

Worksheet 9

The 'SCategory-Input' worksheet (sheet 9) gives the parameters needed for the service categories which are not provided by the market study. These parameters include mean packet size, second moment of packet size, maximum allowable mean packet delay, and maximum allowable blocking probability. The first three parameters are needed for packet-switched service categories SC 11–20 while the last parameter is needed for circuit-switched service categories SC 1–10. The values are given separately for the three different forecast years.

Worksheets 10–13

The results of the market studies from Report ITU-R M.2072 for the three different forecast years are collected in the worksheets 'Market-Input 2010', 'Market-Input 2015' and 'Market-Input 2020' (sheets 10 to 12). Values are given per service category and service environment for user density, session arrival rate per user, mean service bit rate and average session duration. The worksheets show the upper and lower limits from the market studies and the actual values within the range which are selected by defining the percentages in worksheet 'Market-Setting' (sheet 2). The values are given separately for uplink and downlink directions and unicast and multicast traffic. The selection of the actual values is done in the macros in the 'ReadInput' module. The intermediate values for the mobility ratios are calculated on hidden worksheets which can be made visible from Format toolbar by selecting 'Sheet' and 'Unhide' and by choosing the corresponding mobility scenario (lowest, middle or highest) and the corresponding year. The values on 'Market-Studies' worksheet (sheet 13) correspond to the output of Step 3 in the methodology flow chart.

5.4 Intermediate Calculation Steps

Intermediate calculation steps are performed in worksheets 14–26 using the input parameter values from worksheets 2–13. Worksheets 14–16 perform initial calculations on the market parameters, worksheets 17–19 calculate the distribution ratios (Step 4 in methodology flow chart), worksheets 20–23 distribute the traffic according to distribution ratios (Step 4 in methodology flow chart), worksheets 24 and 25 perform the capacity requirement calculations (Step 5 in methodology flow chart), and finally worksheets 26 and 27 calculate the spectrum requirements (Steps 6–9 in methodology flow chart).

Worksheet 14

The 'AreaArrivalRate' worksheet (sheet 14) shows the session arrival rate per area unit expressed in session arrivals/s/km^2. The session arrival rate per area unit is computed as the product of the user density and the session arrival rate per user which are given in 'Market-Studies' worksheet. The values are given for unicast traffic for the 20 service categories in six service environments in three different forecast years separately for uplink and downlink directions.

Worksheet 15

The 'SessionVolume' worksheet (sheet 15) gives the traffic volume of one session in kbit/session. The session volume is computed as the product of the average session duration and the mean service bit rate which are given in 'Market-Studies' worksheet. The values are given for the 20 service categories in six service environments in three different forecast years separately for uplink and downlink directions and unicast and multicast traffic.

Worksheet 16

The 'AreaTrafficVolume' worksheet (sheet 16) gives the traffic volume per area unit expressed in kbit/s/km^2. The traffic volume per area unit is computed as the product of the session arrival rate per area unit from 'AreaArrivalRate' worksheet and the traffic volume of one session from 'SessionVolume' worksheet. The values are given for unicast traffic for the 20 service categories in six service environments in three different forecast years separately for uplink and downlink directions.

Worksheets 17–19

The 'Dist-Ratio-Input' worksheet (sheet 17) starts the distribution functionality which corresponds to Step 4 of the methodology flow chart given in Figure 4.2. The distribution functionality calculates the distribution ratios which are used to divide the total traffic among the radio environments and RATGs. The actual calculations are done with the macros given in the module 'DistributionModule'. The 'Dist-Ratio-Input' worksheet gives the supported radio environments in different service environments in different RATGs and forecast years as well as the supported service categories in different radio environments in different RATGs, service environments and forecast years. The 'Dist-comb' worksheet (sheet 18) gives the possible combinations of service categories, radio environments and RATGs in different service environments and forecast years. Value '1' in the tables denotes a possible combination while value '0' denotes an impossible combination. The distribution is done separately for uplink and downlink directions as well as unicast and multicast traffic and therefore separate tables are provided. The 'Dist-Ratio-Matrix' worksheet (sheet 19) gives the final distribution ratios which are used to split the traffic among the different radio environments and RATGs. A checking function is also performed to find out if all traffic can be distributed. If there are errors in traffic distribution, i.e. part of traffic remains undistributed, the corresponding error is reported in 'Dist-Ratio-Matrix' sheet and a warning message is also written in the 'Warnings' section on the front sheet 'Main'.

Worksheets 20 and 21

After the traffic distribution ratios are calculated, the traffic is distributed among the radio environments and RATGs with the distribution ratios given in 'Dist-Ratio-Matrix' worksheet. The traffic calculation with the distribution functionality presented in Section 4.3.3 is performed with the macros given in the module 'OutputTrafficModule'. The traffic values after traffic distribution are written to 'PSTraffic-op' and 'CSTraffic-op' worksheets for packet-switched and circuit-switched service categories, respectively. The 'PSTraffic-op' worksheet (sheet 20) gives the offered traffic per cell for packet-switched service categories in kbit/s/cell for RATG 1 and RATG 2 in different service environments, radio environments and forecast years. Separate tables are given for uplink and downlink directions. The 'CSTraffic-op' worksheet (sheet 21) gives the offered traffic per cell for circuit-switched service categories in session arrivals/s/cell for RATG 1 and RATG 2 in different service environments, radio environments and forecast years. Separate tables are given for uplink and downlink directions as well as for unicast and multicast traffic.

Worksheets 22 and 23

The traffic in 'CSTraffic-op' and 'PSTraffic-op' worksheets is given per service environment while in a later stage the spectrum requirements are given per teledensity. One teledensity consists of one or more service environments as shown in 'SE-Input' worksheet. Therefore, the traffic is accumulated over the service environments that belong to the same teledensity. The traffic accumulation over service environments is done in 'PSTraffic-op-teledensity' and 'CSTraffic-op-teledensity' worksheets for packet-switched and circuit-switched service categories, respectively. The accumulation is done with the macros given in the module 'OutputTrafficModule'. The 'PSTraffic-op-teledensity' worksheet (sheet 22) gives the offered traffic in kbit/s/cell for packet-switched service categories SC 11–20 in RATG 1 and RATG 2 in different teledensities, radio environments and forecast years. The 'CSTraffic-op-teledensity' worksheet (sheet 23) gives the offered traffic in Erlang/cell as well as the mean service bit rate in kbit/s for circuit-switched service categories SC 1–10 in RATG 1 and RATG 2 in different teledensities, radio environments and forecast years. Separate tables are given for uplink and downlink directions as well as for unicast and multicast traffic.

Worksheet 24

The 'PSCapacity_calculation' worksheet (sheet 24) performs the capacity calculations for packet-switched service categories SC 11–20 according to Section 4.3.5. The actual calculations using the M/G/1 nonpreemptive priority queueing model are done with the macros given in module 'PSCapacityModule'. The capacity calculation gives the required system capacity in kbit/s/cell for all packet-switched service categories for RATG 1 and RATG 2 in different teledensities, radio environments and forecast years. The capacity requirements are calculated separately for uplink and downlink directions.

Worksheet 25

The 'CS-CapacityCalc' worksheet (sheet 25) performs the capacity calculations for circuit-switched service categories SC 1–10 according to Section 4.3.4. The actual calculations using the multidimensional Erlang-B formula are done with the macros given in module 'CSCapacityModule'. The capacity calculation gives the required system capacity in kbit/s/cell for all circuit-switched service categories for RATG 1 and RATG 2 in different teledensities, radio environments and forecast years. The capacity requirements are calculated separately for uplink and downlink directions as well as for unicast and multicast traffic.

Worksheet 26

The 'Spectrum_requirement' worksheet (sheet 26) calculates the spectrum requirements of RATG 1 and RATG 2 in different radio environments, teledensities and forecast years as presented in Section 4.3.6. The capacity requirements from 'PSCapacity_calculation' and 'CS-CapacityCalc' worksheets are combined, and the capacity requirements are divided by the corresponding spectral efficiency values from the 'RATGEff-Input' worksheet. Then the resulting spectrum requirements of unicast and multicast transmission modes are combined.

Worksheet 27

Finally, the last worksheet 'Adjs&AggSpectrum' (sheet 27) adjusts the spectrum requirement with the number of network deployments from the front sheet 'Main', and the spectrum requirements are combined over the radio environments. Also the guard bands between operators from 'RATG1&2Def-Input' worksheet are taken into account. The total spectrum requirements for RATG 1 and RATG 2 are taken as the maximum over the teledensities. The final spectrum requirements are given for RATG 1 and RATG 2 in the three different forecast years.

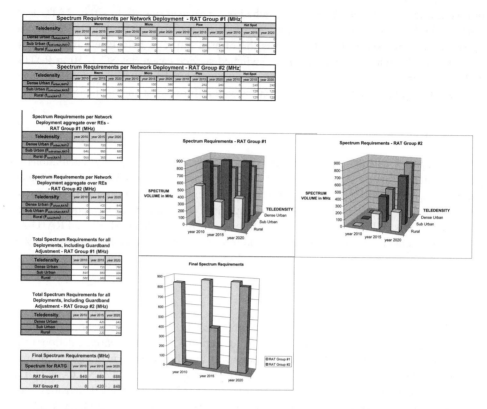

Figure 5.2 Last worksheet of the tool for ITU spectrum calculation methodology.

5.5   Outputs from Software Tool

As output, the MS Excel tool calculates the spectrum requirements of RATG 1 and RATG 2 in the forecast years 2010, 2015 and 2020. The spectrum requirement results are calculated on the last worksheet 'Adjs&AggSpectrum'. The final results are shown on 'Adjs&AggSpectrum' worksheet as well as on the front sheet 'Main'. The last worksheet is shown in Figure 5.2. The last worksheet contains three figures which illustrate the spectrum requirements. The first figure shows the spectrum requirements of RATG 1 in dense urban, sub-urban and rural teledensities in 2010, 2015 and 2020. The second figure shows the same information for RATG 2. The third figure shows the final spectrum requirements of RATG 1 and RATG 2 in three forecast years after taking the maximum over teledensities. The spectrum requirements of the two RATGs are drawn in the same figure for comparison. To estimate the total spectrum requirement of IMT systems, the spectrum requirements of RATG 1 and RATG 2 can be added up.

In addition to the final spectrum requirements of RATG 1 and RATG 2 in the three fore-cast years, the tool also provides time-shifted spectrum requirements of the two RATGs. The time-shifted format defines three development scenarios to characterize the differences in the market development and the deployment of RATGs in different countries. The deployment scenarios include fast, medium and slow deployment. The medium deployment scenario represents the default scenario which is directly obtained as output of the calculations performed in the MS Excel tool. The medium deployment scenario characterizes the average global common market situation. The fast deployment scenario applies to countries where the new systems are introduced fast and the usage of the new services develops fast. The slow deployment scenario corresponds to countries which employ the new systems when they are at mature state.

The time-shifted spectrum requirements for the medium scenario are the same as the final spectrum requirements of RATG 1 and RATG 2 in 2010, 2015 and 2020 presented in the 'Output' section of the front sheet 'Main' as well as in the last worksheet 'Adjs&AggSpectrum'. The spectrum requirements of the fast scenario are obtained by shifting the medium scenario so that the spectrum requirement in medium scenario in 2015 and 2020 corresponds to the spectrum requirement of fast scenario in 2010 and 2015, respectively. The spectrum requirements of the slow scenario are obtained by shifting the medium scenario so that the spectrum requirement in medium scenario in 2010 and 2015 correspond to the spectrum requirement of slow scenario in 2015 and 2020, respectively.

6

Market Data

Marja Matinmikko and Mitsuhiro Azuma

The ITU spectrum requirement calculation methodology for the future development of IMT-2000 and IMT-Advanced, presented in Chapter 4, includes several input parameters for which unique values are needed. An important part of the spectrum requirement calculation is the predicted demand of future services in the time span 2010 to 2020. The ITU market study entitled 'World Mobile Telecommunication Market Forecast' provided in Report ITU-R M.2072 includes a comprehensive study of market forecasts for future wireless services from different parts of the world.

This chapter describes the market-related input parameters for the spectrum calculation methodology including the following parameters:

- user density

- session arrival rate per user

- average session duration

- mean service bit rate

- mobility ratio.

The approach for collecting the market data and the key findings of the future mobile market given in Report ITU-R M.2072 are explained in Section 6.1. Section 6.2 describes the individual market study parameters and characterizes their meaning and role in the spectrum calculation methodology. The ITU market study combines market forecasts from different countries and gives ranges of values for the different market study parameters. Moreover, the individual market studies characterize future services on service level while the spectrum requirement calculation methodology uses the notion of service categories which may consist of several services. Therefore, Section 6.3 describes the approach to analyze the market study

Spectrum Requirement Planning in Wireless Communications   Edited by Hideaki Takagi and Bernhard H. Walke
© 2008 John Wiley & Sons, Ltd

data to be used as input to the spectrum calculation methodology. Finally, an example set of unique values for the market study parameters is given in Section 6.4 to be used in the numerical example calculations in Chapter 8.

6.1   Collection of Market Data

The information needed to forecast the evolution of mobile market and services for the future development of IMT-2000 and IMT-Advanced was gathered by means of a questionnaire which was distributed worldwide. The ITU questionnaire included in Administrative Circular Letter CACE/326 was sent out in 2004 to obtain quantitative information on the future mobile market from different organizations including organizations outside the ITU. The responses to the questionnaire including information from about thirty administrations and organizations were collected into Report ITU-R M.2072 which was completed in 2005. The questions in the ITU questionnaire are summarized in Section 6.1.1 and the key findings based on Report ITU-R M.2072 are summarized in Sections 6.1.2 and 6.1.3.

6.1.1   Questionnaire on services and market

Annex 1 of Report ITU-R M.2072 shows the 'Questionnaire on the services and market for the future development of IMT-2000 and IMT-Advanced' which was sent out in Administrative Circular Letter CACE/326 from ITU-R (http://www.itu.int/md/R00-CACE-CIR-0326/en). It includes the following questions:

Q.1   Services and market survey for existing mobile services.

This question requests to list the information on the following parameters as they existed at the end of the year 2003 regarding 2G and IMT-2000 mobile communication services and market.

- number of subscribers
- traffic volume
- expected time frame for transition from 2G to IMT-2000
- regional and geographical characteristics
- system type
- area covered
- commencement of operation
- economics of network deployment.

Q.2   Key market parameters.

This question requests to list the key parameters and issues that each organization considers necessary for the market size estimation, and comment on the impact of each parameter and issue on the market. The market size of the future development of IMT-2000 and IMT-Advanced can be estimated from parameters such as density of potential users, call/session duration, service bandwidth and so forth. It is also necessary for the market size estimation to take such key issues into account as services usage and usage environments.

For instance, the market study may provide the following parameters:

- number of mobile voice service subscribers;
- traffic volume for data service from mobile subscribers;
- density of potential users in urban area during a specific time of day;
- density of potential users in a rural/remote area for specific time of day;
- average call duration of voice service during a specific time of day;
- session activation rate of web browsing service by heavy users;
- services and related QoS parameters: guaranteed/best-effort, real time/non-real time, fixed/variable bit rate;
- usage environments: high/low mobility, indoor/outdoor, business/private and urban/non-urban.

The unit of measuring the above parameters may be as follows:

- number of service requests per unit time and area (sessions/s/m^2).
  The value of this parameter could be provided for appropriate time intervals during the day, and separately for working days and non-working days, in order to account for non-simultaneous occurrence of the peak offered traffic of different service types;
- percentage of users in a set of mobility classes: stationary, pedestrian and vehicular;
- maximum required mobility support (km/h);
- average service duration (s) or the amount of data;
- required minimum, average and maximum bit rate measured at the application layer (bit/s) for uplink and downlink directions;
- required end-to-end delay (s).

Q.3 Service and market forecast for future development of IMT-2000 and IMT-Advanced.

This question requests to provide forecasts on the future status of the key parameters listed in Q.2 regarding services and markets of the future development of IMT-2000 and IMT-Advanced. Future trends of the parameters are to be described in this question. It is also requested to mention the evolution of service capabilities.

Q.3.1 Service issues.

This requests to describe the applications envisaged for the future development of IMT-2000 and IMT-Advanced, which may be pervasive from the year 2010 to 2020. Examples of potential future applications are listed in Section 6.1.2.

Q.3.2 Market issues.

This requests to describe the trends and the scale of the market (quantitative information desired) related to mobile communication from the year 2010 to 2020, including the statement on how this information was obtained (e.g. methods and parameter values).

Q.3.3 Preliminary traffic forecast.

This requests to provide information related to service traffic to be provided by the future development of IMT-2000 and IMT-Advanced from the year 2010 to 2020. The volume of traffic may be derived from information such as the number of service subscribers, service activation rate, service duration, transferred data size taking into account affordable expenses of the subscribers for this service. It may also be derived from information such as uplink and downlink speed, traffic class, service environment and economics of network deployment. It is preferred to receive the information related to the traffic volume for each service in the answer to Q.3.1.

Q.3.4 Related information.

Any relevant information is requested such as:

- impacts of the number of operators;
- peak hour/peak ratio of traffic volume;
- expected affordable expenses per subscriber for the services on the average;
- economics of network deployment.

Q.4 Service and market forecast for other radio systems.

The future development of IMT-2000 and IMT-Advanced are envisaged to go along with other radio systems such as wireless LAN and broadcasting systems. This question requests to list any radio systems that might interwork with the future development of IMT-2000 and IMT-Advanced as well as to forecast the future status of the parameters from Q.3. This also requests to indicate the percentage of users who subscribe to several systems/operators.

Q.5 Driving forces of the future market.

This question requests to list any items which will be the driving forces in the markets of the future development of IMT-2000 and IMT-Advanced, and to estimate their impacts and timing. There may be different drivers for different market areas, e.g. urban versus rural/remote, in respective countries.

Q.6 Any other views on future services.

This question asks if there are any other views on the services to be provided by the future development of IMT-2000 and IMT-Advanced which are not described in Recommendation ITU-R M.1645. If so, an elaborate description is requested.

6.1.2 Example of envisaged applications

In response to the questionnaire shown in the preceding section, a large number of applications/services were envisaged for the future development of IMT-2000 and IMT-Advanced. In the following, we list representative applications taken from Report ITU-R M.2072 with our brief comments. This listing corresponds to Step 1 in the analysis process for collected market data described in Section 6.3.2 and shown in Figure 6.1 below. At this moment, the applications are listed at random. However, we have classified them according to the direction of transmitting application data shown in Figure 1.1 in Chapter 1. The applications from responses to the questionnaire are to be grouped into service categories in Step 4 of the market data analysis in Figure 6.1.

- Human-to-human communication

    - voice telephony
    Traditional voice telephony will continue to play an important role in the future.

    - voice over Internet protocol (VoIP)
    Telephony over Internet includes services such as Skype.

    - video telephony
    Video telephony will introduce more attractive services than voice telephony.

    - video conference
    This service can be provided by the technology of video telephony.

    - low-priority e-mail
    Messaging services will also continue to play an important role in the future.

    - photo mail/video mail
    With the advent of camera functionality on mobile terminals and the introduction of flat rate high-speed services, still photo mail and video mail services will attract users more quickly than ever.

    - interactive gaming
    Interactive gaming, such as multiplayer gaming, will increase with introduction of high-performance mobile devices and adoption of broadband technology.

    - telemedicine
    The patients in the rural area can obtain equable medical care which will spread wide into every corner of the country.

    - collaborative work
    This includes multimedia information exchange, file and application sharing.

- Machine-to-human communication

    - mobile Internet
    Internet services on mobile terminal with the almost same quality as on PC are available all over the place such as wireless hot spots.

    - web browsing
    This service, now commonly used by PC, will be accelerated on mobile terminals with the advent of flat rate service for packet access.

    - media downloading
    Contents downloading services via mobile terminals will become popular such as e-newspaper, e-book, music and game data.

    - video streaming (entertainment)
    With the dissemination of broadband services, video streaming services over mobile terminals will increase.

- mobile TV and video

  With the convergence of communication and broadcasting technologies, TV and video services, such as 'One Segment', are available on mobile terminals.

- e-learning

  This may become popular to meet trainees' demand for time and cost saving for education. This includes conversational, video streaming and background services.

- Intelligent Transport System (ITS) for navigation

  Car navigation systems with telecommunications capability are emerging, which can afford surrounding information including maps, car parks, restaurants, malls, shops, etc.

- emergency notification/disaster prediction

  Earthquake and tsunami prediction/notification are important communications. The emergency rescue, which requires immediate preparatory care for a person, is also a very important service.

- exercise monitor and instruction

  To keep health condition, exercise occasions are increasing and become popular. An exercise monitor management system is able to give instructions by analyzing uploaded bio-medical or physical data.

- Human-to-machine communication

  - video uploading

    Personal video uploading, such as You-Tube, is becoming popular.

  - health care/health check, remote diagnostics, and medication information, medical data provision

    With the advent of digital home appliance or home networks, new services around home environment are under development such as medical/health care/surveillance monitoring applications for elderly people in their home environments.

- Machine-to-machine communication

  - peer-to-peer (P2P) file transfer

    With the dissemination of broadband services, P2P file exchange services over Internet and P2P applications over mobile terminals will be increasing.

  - telemetering

    Telemetering is one of the applications using machine-to-machine telecommunications. These include telemetering of earthquakes, heavy rain and wind, temperature and humidity. For home applications, telemetering of utilities such as electricity, gas, water and sewer are envisaged. For public applications, Automatic Vehicle Monitoring (AVM) and quality monitoring of water supply are envisaged.

– ITS probe

The position location service is necessary on emergency cases for security reasons. The reporting capability is required on various types of equipment and devices.

6.1.3 Overview of future mobile telecommunication market

The responses to the questionnaire on services and market described in Section 6.1.1 are summarized in Report ITU-R M.2072 which depicts the trends and evolution of the future mobile telecommunication market. This section summarizes the findings of the market report.

Question Q.1 requests to characterize the market situation of 2G and IMT-2000 in the year 2003 which is important background information for the prediction of the future mobile market. In a global view, the turning point in the transition from 2G to IMT-2000 was the year 2004 when new tariffs and enhanced handsets were introduced. The transition was expected to reach its peak in 2006 and then to decrease slowly in favor of IMT-2000. The regional and national market situations in 2003 were very different in different parts of the world in terms of numbers of users, traffic volumes, penetration rates and transition from 2G to IMT-2000. While some countries, such as Japan and Korea, enjoyed high numbers of users and good progress in transition to IMT-2000 already in 2003, some other countries showed much lower overall penetration rates for the mobile services.

In response to Questions Q.2 and Q.3 on the key market parameters and forecasts for the future development of IMT-2000 and IMT-Advanced, Report ITU-R M.2072 provides summaries of several case studies on the detailed market forecasts. The report lists the key market issues that influence the market forecast. The key issues include, for example, the social and economical aspect (e.g. population and income), technological aspect, market aspect, service aspect (e.g. tariffs and device cost), traffic volumes and types, regulatory aspect and industrial aspect. The report also shows traffic forecasts in figures. The traffic forecasts predict a strong growth in the mobile market. The forecasted values for the five market study parameters used in the spectrum calculation methodology are given as ranges since the different case studies are combined into a single market prediction.

Question Q.4 requests to provide market forecast for radio systems other than the future development of IMT-2000 and IMT-Advanced. In the future, IMT-Advanced systems and other radio systems will be integrated into a large system which consists of complementary elements. Moreover, it is expected that almost all mobile terminals will support multiple radio access capabilities. In this context, the percentages of users of the different radio access technologies will not be constant. The market forecasts for the other radio systems included in Report ITU-R M.2072 show a strong growth in the numbers of subscribers.

The driving forces in the market of the future development of IMT-2000 and IMT-Advanced (Question Q.5) are identified in Report ITU-R M.2072 together with the estimates of their impact and timing. The driving forces are classified into technology, market, regulatory and idealistic forces. The technology forces include, for example, broadband, VoIP, mobile voice communication, distributed computing, target services, amalgamation of technologies and network reconfiguration. In the future, the quality and speed of mobile services should be equivalent to that of wired broadband networks.

Market forces include economic development, ubiquitous access to Internet, changing business model, group collaboration, feature transparency with landline, carrier roaming between high-speed data networks, pricing model, and complex business models. In the end the economic development determines the consumption capabilities and therefore it is important to consider socio-economic factors and consumer demand. Moreover, the business models for the future mobile market will be more complex than they are today.

Regulatory forces include policies upon carriers, unlicensed technologies and spectrum management as well as cooperation and competition among network providers. Efficient spectrum management will be a key issue in the future systems and flexible spectrum use will play a big role. Moreover, today's competitive coexistence of network providers should be replaced by willingness to cooperate.

Idealistic forces consider the relationships of concepts and challenge the traditional thinking. Idealistic forces include, for example, communication systems versus communication services, service delivery to terminal versus service delivery to user, storage versus communication, data source versus data sink, streaming versus download/upload, and local versus remote versus distributed processing.

6.2   Use of Market Parameters in the Methodology

The ITU spectrum calculation methodology uses a limited set of market study parameters to characterize the future wireless services. The market study parameters characterize the demand of twenty different service categories in six service environments in three forecast years. The parameter values are given separately for uplink and downlink transmission directions as well as for unicast and multicast traffic. In the following, the meaning and role of the different market study parameters used in the methodology are explained.

6.2.1   User density

The user density $U_{m,n}$, expressed in users/km^2, presents the average density of potential users of the different service categories in the different service environments and forecast years. The user density parameter is used in Step 2 to Step 4 of the methodology flow chart in Figure 4.2 where the market data is analyzed, the traffic is distributed among the different radio environments and RATGs, and the traffic per cell is calculated. For unicast service categories, the traffic calculation, shown in Equation (4.13), presents the session arrival rate per cell in sessions/s/cell as the product of the distribution ratio, the user density, the session arrival rate per user, and the cell area. The product of the user density with the corresponding cell area gives the average number of active users in a cell. The session arrival rate per user shows the rate at which the users, on average, request the service. The larger the user density, the more users there are in a given cell and the more traffic is being generated, leading to the larger unadjusted spectrum requirements. The unadjusted spectrum requirements correspond to the output from Equation (4.40).

The user density parameter is given per service environment but the parameter values do not differ between the different radio environments. In general, the user densities may be different in different cell types and in particular the smaller cells are more densely populated. This fact is taken into account in the traffic distribution part in Step 4 of the methodology flow chart in Figure 4.2, where the smaller cells are favored over the larger cells in allocating the user traffic.

6.2.2   Session arrival rate per user

The session arrival rate per user $Q_{m,n}$, expressed in sessions/s/user, characterizes the rate of demand for different service categories in different service environments and forecast years per user. The session arrival rate per user is used in Equation (4.13) for unicast traffic, where the session arrival rate per cell is computed as the product of the distribution ratio, the user density, the session arrival rate per user, and the cell area. Therefore the session arrival rate per user and the user density have similar effects in the spectrum calculation methodology. An increase in the session arrival rate per user leads to an increase in the traffic volume, which in general leads to an increase in the unadjusted spectrum requirements.

6.2.3   Average session duration

The average session duration $\mu_{m,n}$, expressed in s/session, gives the average duration of one session of the different service categories in the different service environments and forecast years.

The product of the average session duration with the session arrival rate per cell gives the offered traffic per cell in Erlang/cell as shown in Equation (4.16). This quantity is further used in the capacity calculation of circuit-switched service categories. In general, an increase in the average session duration leads to an increase in the traffic volumes, which may lead to larger unadjusted spectrum requirements.

The product of the average session duration and the mean service bit rate gives the traffic volume of one session. When the traffic volume of one session is multiplied with the session arrival rate per cell, the offered traffic volume per cell in bit/s/cell is obtained in Equation (4.18). This traffic volume per cell is further used in the capacity calculation of packet-switched service categories.

6.2.4   Mean service bit rate

The mean service bit rate $r_{m,n}$, expressed in bit/s, characterizes the bit rate requirement for the different service categories in the different service environments and forecast years. The mean service bit rate parameter is used in three different instances in Step 4 and Step 5 of the methodology flow chart in Figure 4.2.

First, the mean service bit rate of a service category is used to find out which service categories can be supported by the different RATGs in different radio environments as explained in Section 4.3.2. This is done by comparing the mean service bit rate requirement of a service category in different service environments with the application data rate of different RATGs in different radio environments. A service category can be supported in a given radio environment of a given RATG only if the application data rate provided by the RATG in the given radio environment exceeds the mean service bit rate required by the service category. When the mean service bit rate is increased, the availability of radio environments and RATGs may be reduced, which leads to different traffic volumes being distributed to radio environments and RATGs. Therefore, a change in the mean service bit rate can lead to changes in the spectrum requirements, which cannot be simply predicted from the individual changes in the values. In general, if the value of the mean service bit rate is in the vicinity of the value of the application data rate parameter, then even a small change can result in changes in the spectrum requirements if the availability of radio environments and RATGs is changed by the corresponding change in the mean service bit rate.

Second, the mean service bit rate is used in the capacity calculation for circuit-switched service categories. The offered traffic volume per cell for circuit-switched service categories is expressed in Erlang/cell as shown in Equation (4.16). This traffic volume does not consider the mean service bit rate, but the capacity calculation does take into account the mean service bit rate which is calculated in Equation (4.17).

Third, the mean service bit rate is used to calculate the traffic volumes for packet-switched service categories. The multiplication of the mean service bit rate and the average session duration gives the traffic volume of one session. When the traffic volume of one session is multiplied with the session arrival rate per cell, the offered traffic volume per cell in bit/s/cell is obtained in Equation (4.18). Therefore an increase in the mean service bit rate leads to an increase in the traffic volume per cell for packet-switched service categories, which may lead to a larger unadjusted spectrum requirement.

6.2.5   Mobility ratios

The mobility ratios describe the speed of movement of the users. The traffic in each service category is divided into different mobility classes to predict the user behavior in terms of what are the likely usage situations for the different service categories. The sum of the mobility ratios of a given service category in a given service environment and forecast year is equal to one, meaning that all traffic is associated with corresponding mobility ratios.

The mobility ratios used in the ITU market studies provided in Report ITU-R M.2072 include stationary, low, high and super-high mobilities. The ITU spectrum calculation methodology presented in Recommendation ITU-R M.1768 defines stationary/pedestrian, low and high mobilities. Therefore the mobility ratios from the market studies must be pre-processed, and this is done in the MS Excel tool described in Chapter 5. The adjustment of the market study mobility ratios into the mobility ratios used in the methodology is done according to the principle explained in Section 4.3.1. An input parameter J_m is used in the adjustment of the mobility ratios in service environment m. The sum of high and super-high mobility classes from the market study is divided into the low and high mobility classes of the methodology using the J_m parameter, which gives the fraction of low mobility class in the methodology. The detailed equations are given in Section 6.3.6 with the help of Figure 6.5.

Mobility ratios are used in the traffic distribution algorithm in Step 4 of the methodology flow chart in Figure 4.2 where the traffic distribution ratios among the radio environments are calculated. The mapping between mobility classes and radio environments is considered in the traffic distribution, as explained in Section 4.3.2. High mobility is supported only in macro cells, while users with low mobility can be supported in both macro and micro cells. Stationary and pedestrian users can be supported in all radio environments.

The J_m parameter is an input parameter for the calculations. When the J_m parameter is decreased, a larger portion of traffic is in the high mobility class, which results in more traffic being distributed to macro cells because only macro cells can support high velocities.

Since the market study in Report ITU-R M.2072 includes ranges of values for mobility ratios, for actual calculations the user must select individual values within the ranges. The selection of mobility ratios is done by choosing mobility scenarios from the three alternatives (1 = lowest, 2 = middle, 3 = highest) in the 'Market-Setting' worksheet (sheet 2) of the MS Excel tool as explained in Section 5.3. By selecting higher mobility scenarios, it is likely that more traffic is distributed to macro cells because only macro cells can support the high velocities. By selecting lower mobility scenarios, it is likely that more traffic is distributed to hot spots and pico cells. The changes in the mobility scenarios lead to changes in the traffic distribution which in turn can change the traffic volumes and thus the unadjusted spectrum requirements.

6.3   Analysis of Collected Market Data

The responses to the questionnaires in the service view document are processed to obtain the values for the market study parameters described in Section 6.2, which are needed as input to the spectrum calculation methodology. This section describes a method presented in Report ITU-R M.2072 to analyze the market data from various responses to the questionnaires. The market data analysis uses the notion of applications and services. A service is the basic element which builds up the applications. An application is a higher level definition to categorize the collected services. An application may consist of several services that occur independently.

6.3.1   General process

The general process for the market data analysis is shown in Figure 6.1. Step 1 is to list all different applications and services. Step 2 is to specify traffic attribute parameter values of each service. The traffic attribute parameter values are market study parameters that are related to the traffic characteristics of the service. Step 3 is to specify market attribute parameter values of each service. The market attribute parameter values are market study parameters that are related to the users' perspective. Step 4 is to map the services into service categories in different service environments. Finally, Step 5 is to calculate the values for the different market study parameters for each service category in each service environment. The different steps are described in more detail in the following Sections 6.3.2–6.3.6.

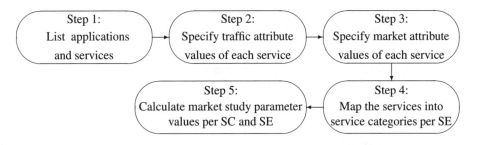

Figure 6.1 General process for the market data analysis.

6.3.2   List applications and services

Step 1 in the market data analysis in Figure 6.1 is to list all foreseeable future wireless applications and services. The listed services should be general and essential enough to categorize all the collected services concisely and appropriately. The lists of all applications and services are first identified and written in the first and second columns of Figure 6.2, respectively. The obtained application list shown in this figure should be categorized by considering their traffic and market attributes in later steps. Individual responses to the questionnaires are analyzed in terms of whether they include a list of applications and services. When the response includes such a list, every item is merged in the list of applications and services as in Figure 6.2. This figure includes two example applications, i.e. 'existing applications' and 'town monitoring systems' and 11 example services associated with the two applications.

6.3.3   Specify traffic attribute values for services

In Step 2 of the market data analysis in Figure 6.1, the values of the traffic attribute parameters are specified based on the lists of applications and services developed in Step 1. The traffic attribute parameters include the average session duration per service and the mean service bit rate in different service environments. Typical values for these two market study parameters are collected into the third and fourth columns of Figure 6.2, respectively. These values can be used to assist in deriving the applications and services from the collected market data. This figure shows example values for the mean service bit rate for the two example applications and the 11 example services associated with these applications.

6.3.4   Specify market attribute values for services

In Step 3 of the market data analysis in Figure 6.1, the values of the market attribute parameters are defined for each service in each service environment. The market attribute parameters include the user density, the session arrival rate per user and the mobility ratios. Figure 6.3 shows an example application and services with their traffic and market attributes, including two example services associated with the application 'town monitoring systems'. Figure 6.2 is utilized for input to Step 3, and the result of this step is merged into Figure 6.3 that may already include some results from the response to the questionnaire on services and market. When multiple applications are related to the same service with different traffic/market attribute values, the range of those values should be given in Figure 6.3.

6.3.5   Map services into service categories

In Step 4 of the market data analysis in Figure 6.1, the services are mapped into service categories. Each service is characterized by a service type and a traffic class as exemplified in Figure 6.4. Similarly, the service categories are built up of combinations of service types and traffic classes as explained in Section 4.2.1. All the services listed in Figure 6.3 are then mapped into Figure 6.4 based on the combinations of service types and traffic classes. As a result, all services are associated with the corresponding service categories. The mapping can be done separately in each service environment.

Applications	Services		Traffic attributes	
			Average session duration	Mean service bit rate
Existing applications	Voice (multimedia and low rate data/ conversational)			64 kbit/s
	Video phone (medium multimedia/ conversational)			384 kbit/s
	Packet	E-mail (very low rate data/background)		1 kbit/s
		Video mail (medium multimedia/background)		512 kbit/s
		Mobile broadcasting (high multimedia/streaming)		5 Mbit/s
		Internet access (high multimedia/background)		10 Mbit/s
Town monitoring systems	Voice (multimedia and low rate data/ conversational)			64 kbit/s
	Video communication (medium multimedia/conversational)			384 kbit/s
	Medium rate data transmission for town information monitoring (super-high multimedia/interactive)			384 kbit/s
	Low rate data transmission for reservation of restaurants, etc. (very low rate data/interactive)			1 kbit/s
	File transfer (super-high multimedia/ background)			50 Mbit/s

Figure 6.2 Example applications and services with their traffic attributes.

Figure 6.4 shows example mappings of the services listed in Figure 6.2 into the service categories defined in Section 4.2.1. The services voice, video phone, e-mail, video mail, mobile broadcasting and Internet access associated with the application 'existing applications' in Figure 6.2 are mapped into the appropriate combinations of service types and traffic classes in Figure 6.4. Similarly, the services associated with the application 'town monitoring systems' including voice, video communication, medium data rate transmission for town information monitoring, low rate data transmission for reservation of restaurants, and file transfer in Figure 6.2 are mapped into the appropriate service categories in Figure 6.4.

Application	Services s	SC n	SE m	Traffic and market attributes				Mobility ratio $\mathrm{MR}_{m,s}^{\mathrm{Market}}$ (%)			
				$U_{m,s}{}^{a}$	$Q_{m,s}{}^{b}$	$\mu_{m,s}{}^{c}$	$r_{m,s}{}^{d}$	Sta-tionary	Low	High	Super-high
Town monitoring systems	Voice $s = 1$		1								
			2								
	Video communication $s = 2$										

[a] User density $U_{m,s}$ (users/km^2).
[b] Session arrival rate per user $Q_{m,s}$ (sessions/s/user).
[c] Average session duration $\mu_{m,s}$ (s/session).
[d] Mean service bit rate $r_{m,s}$ (bit/s).

Figure 6.3 Example applications and services with their traffic and market attributes.

6.3.6   Calculate market study parameter values for input to methodology

After defining the mapping of services into service categories, the actual values for the market study parameters per service category are calculated in Step 5 of the market data analysis in Figure 6.1. The actual calculations of the market study parameters are done with the following equations.

User density

The user density $U_{m,n}$ (users/km^2) of service category n in service environment m is calculated as the summation of the user densities of each service mapped into this service category according to

$$U_{m,n} = \sum_{s \in n} U_{m,s}, \qquad (6.1)$$

where $U_{m,s}$ (users/km^2) denotes the user density of service s inside service category n in service environment m.

Session arrival rate per user

The session arrival rate per user $Q_{m,n}$ (sessions/s/user) of service category n in service environment m is the weighted average of session arrival rate per user of each service mapped into this service category. The weight of each service is the corresponding user density of the

Traffic class / Service type	Conversational	Streaming	Interactive	Background
Super-high multimedia				File transfer
High multimedia		Mobile broadcasting		Internet access
Medium multimedia	Video phone Video communication		Medium rate data transmission for town information monitoring	Video mail
Low rate data and low multimedia	Voice (existing applications) Voice (town monitoring systems)			
Very low rate data			Low rate data transmission for restaurant reservation	E-mail

Figure 6.4 Mapping of services into service categories.

service, i.e. $U_{m,s}$. The session arrival rate per user is then calculated from

$$Q_{m,n} = \frac{\sum_{s \in n} U_{m,s} Q_{m,s}}{U_{m,n}}, \qquad (6.2)$$

where $Q_{m,s}$ (sessions/s/user) denotes the session arrival rate per user of service s inside service category n in service environment m.

Average session duration

The average session duration $\mu_{m,n}$ (s/session) of service category n in service environment m is the weighted average of average session duration of each service mapped into this service category. The weight is the session arrival rate per area unit, i.e. the product of the user density ($U_{m,s}$) and the session arrival rate per user ($Q_{m,s}$). The average session duration is calculated from

$$\mu_{m,n} = \sum_{s \in n} w_{m,s} \mu_{m,s}, \qquad (6.3)$$

where $\mu_{m,s}$ (s/session) denotes the average session duration of service s inside service category n in service environment m. The weight $w_{m,s}$ (dimensionless) is obtained from

$$w_{m,s} = \frac{U_{m,s} Q_{m,s}}{U_{m,n} Q_{m,n}}, \qquad s \in n. \qquad (6.4)$$

Mean service bit rate

The mean service bit rate $r_{m,n}$ (bit/s) of service category n in service environment m is the weighted average of the mean service bit rate of each service mapped into this service category. The weight is the traffic volume per area unit, i.e. the product of the session arrival rate per area unit ($U_{m,s} Q_{m,s}$) and the average session duration ($\mu_{m,s}$). The mean service bit rate is calculated from

$$r_{m,n} = \sum_{s \in n} \bar{w}_{m,s} r_{m,s},\tag{6.5}$$

where $r_{m,s}$ (bit/s) denotes the mean service bit rate of service s inside service category n in service environment m. The weight $\bar{w}_{m,s}$ (dimensionless) is obtained from

$$\bar{w}_{m,s} = \frac{U_{m,s} Q_{m,s} \mu_{m,s}}{U_{m,n} Q_{m,n} \mu_{m,n}}, \quad s \in n.\tag{6.6}$$

Mobility ratios

The calculation of the mobility ratios involves two stages.

In the first stage, the market study mobility ratio $\text{MR}_{m,n}^{\text{Market}}$ (dimensionless) of service category n in service environment m is calculated as the weighted average of the mobility ratios $\text{MR}_{m,s}^{\text{Market}}$ (dimensionless) of each service s inside service category n in the same service environment m. The weight is the traffic volume per area unit as in the case of the mean service bit rate. Thus the mobility ratios are calculated from

$$\text{MR}_{m,n}^{\text{Market}} = \sum_{s \in n} \bar{w}_{m,s} \text{MR}_{m,s}^{\text{Market}},\tag{6.7}$$

where the weight $\bar{w}_{m,s}$ is given in Equation (6.6). This equation can be applied to all mobility classes, i.e. stationary, low, high and super-high classes for each service category and service environment.

In the second stage, the market study mobility ratios per service category and service environment obtained from Equation (6.7) for stationary, low, high and super-high mobility classes need to be mapped into the methodology mobility ratios for stationary/pedestrian, low and high mobility classes. The mapping is done with the *splitting factor* J_m as explained in Sections 4.3.1 and 6.2.5. The reader is referred to Figure 6.5. Thus the mobility ratio for the stationary/pedestrian mobility class of service category n in service environment m in the spectrum calculation methodology is obtained from

$$\text{MR}_{\text{stat/ped};m,n} = \text{MR}_{\text{stat};m,n}^{\text{Market}} + \text{MR}_{\text{low};m,n}^{\text{Market}}.\tag{6.8}$$

The mobility ratio for the low mobility class of service category n in service environment m in the spectrum calculation methodology is obtained from

$$\text{MR}_{\text{low};m,n} = J_m \text{MR}_{\text{high};m,n}^{\text{Market}}.\tag{6.9}$$

Finally, the mobility ratio for the high mobility class of service category n in service environment m in the spectrum calculation methodology is obtained from

$$\text{MR}_{\text{high};m,n} = (1 - J_m)\text{MR}_{\text{high};m,n}^{\text{Market}} + \text{MR}_{\text{s-high};m,n}^{\text{Market}}.\tag{6.10}$$

Methodology	Market study	
Stationary/pedestrian	Stationary	
	Low	
Low	J_m	High
High	$1 - J_m$	
	Super-high	

Figure 6.5 Mapping of mobility classes from market study to methodology.

Hence the following relation should be clear:

$$MR_{\text{stat/ped};m,n} + MR_{\text{low};m,n} + MR_{\text{high};m,n}$$

$$= MR_{\text{stat};m,n}^{\text{Market}} + MR_{\text{low};m,n}^{\text{Market}} + MR_{\text{high};m,n}^{\text{Market}} + MR_{\text{s-high};m,n}^{\text{Market}}$$

$$= 1 \tag{6.11}$$

for each service category n and service environment m.

6.4 Example Input Market Parameter Value Set

The ITU market study presented in Report ITU-R M.2072 combines the market forecasts from different countries, and gives ranges of values for the different market study parameters described in Section 6.2. The spectrum requirement calculation methodology requires specific values for the market parameters, and thus for input to the methodology, specific values within the ranges must be selected. The actual values for the market study parameters eventually used in the ITU spectrum requirement calculations for WRC-07 are given in Report ITU-R M.2078. The ITU parameter set provided in Report ITU-R M.2078 considers two user density scenarios: lower user density and higher user density. In the following, an example set of market study parameter values is presented based on the higher user density setting in the year 2020 from Report ITU-R M.2078. This example parameter value set is further used in the numerical example calculations in Chapter 8.

The actual values for the user density, session arrival rate per user, average session duration and mean service bit rate parameters are chosen by using the percentage representation in the 'Market-Setting' worksheet (sheet 2) of the MS Excel tool as described in Section 5.3. One can select individual values within the ranges by defining percentages for each service category in each forecast year. The percentage '0%' represents the lower limit of the range while '100%' denotes the upper limit of the range. An example of the percentage

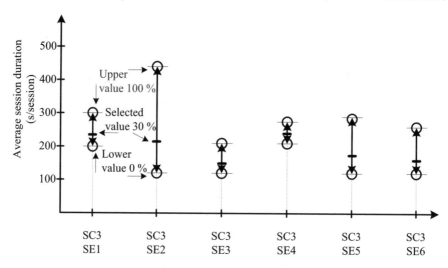

Figure 6.6 Percentage representation for market study parameters.

representation for the market study parameters is shown in Figure 6.6, where the ranged values for the average session duration of the downlink direction of service category (SC) 3 are shown. The percentage '30%' is selected and the corresponding values of average session duration of SC 3 are shown in the different service environments (SEs) 1–6.

Table 6.1 Example J_m factor values for service environment m.

SE 1	SE 2	SE 3	SE 4	SE 5	SE 6
1	1	1	1	0.5	0

The example splitting factor J_m values for converting the mobility ratios from the market studies into the mobility ratios used in the methodology are shown in Table 6.1. In the MS Excel tool, the values of Table 6.1 are input to hidden worksheets performing the intermediate calculations of the mobility ratios. The hidden worksheets can be made visible from the Format toolbar by selecting 'Sheet' and 'Unhide', and by choosing the corresponding mobility scenario (lowest, middle or highest) and the year 2020.

Table 6.2 gives the example market settings in the year 2020 based on the ITU higher user density scenario (taken from Table 29a of Report ITU-R M.2078). Percentage values are defined for user density U, session arrival rate per user Q, average session duration μ and mean service bit rate r. For the mobility ratios, one scenario out of the three alternatives (1 = lowest, 2 = middle and 3 = highest mobility scenario) is selected for each service category. The text 'No range M.2072' in the mobility ratio column denotes that the corresponding service category does not include a range, but instead a single value is given or there is no traffic in the market study in the Report ITU-R M.2072. In such a case, the percentage values do not have a meaning because the single value provided in the market study will be

Table 6.2 Example market setting in higher user density scenario in the year 2020.

SC n	U (%)	Q (%)	μ (%)	r (%)	Mobility ratio
1	25	30	30	30	2 (No range M.2072)
2	25	30	30	30	2
3	25	30	30	30	2
4	25	30	30	30	2
5	25	30	30	30	2
6	25	30	30	30	2 (No range M.2072)
7	25	30	30	30	2
8	25	30	30	30	2
9	25	30	30	30	2 (No range M.2072)
10	25	30	30	30	2
11	25	30	30	30	1
12	25	30	30	30	2
13	25	30	30	30	2
14	25	30	30	30	2
15	25	30	30	30	2
16	25	30	30	30	2 (No range M.2072)
17	25	30	30	30	2
18	25	30	30	30	2 (No range M.2072)
19	25	30	30	30	2 (No range M.2072)
20	25	30	30	30	2

used in the calculations. In the MS Excel tool the values of Table 6.2 are the input to the 'Market-Setting' worksheet. The same market settings are used in all service environments.

The actual values corresponding to the selected market attributes are given in Appendix B. The market study parameter values are given separately for uplink and downlink directions as well as for unicast and multicast traffic.

7

Radio-Related Input Parameters

Marja Matinmikko, Pekka Ojanen and Jussi Ojala

The ITU spectrum calculation methodology for the future development of IMT-2000 and IMT-Advanced includes input parameters which characterize the radio technology-related issues of existing, emerging and future wireless systems. Radio-related issues about IMT-Advanced for the spectrum calculation methodology are presented in Report ITU-R M.2074, where also ranged values for the radio-related parameters are proposed with some justification. The actual values for the radio-related parameters eventually used in the ITU spectrum requirement calculations are given in Report ITU-R M.2078.

A primary requirement for the spectrum calculation methodology is to be technology neutral and generic, and therefore the number of radio-related parameters is kept low. Hence the individual parameters are presented on the radio access technique (RAT) group level instead of considering individual radio access techniques separately. The RAT group (RATG) approach is explained in Section 7.1. In Section 7.2, the role and use of the different radio-related parameters in the spectrum calculation methodology are explained. Finally, Section 7.3 presents an example set of values for the radio-related input parameters. These parameter values are further used in the numerical calculations presented in Chapter 8.

7.1 RAT Group Approach

Agenda Item 1.4 of the World Radiocommunication Conference in 2007 (WRC-07) states the following assignment:

> to consider frequency related matters for the future development of IMT-2000 and systems beyond IMT-2000 taking into account the results of ITU-R studies in accordance with Resolution 228 (Rev. WRC-03),

where we note that 'systems beyond IMT-2000' is now named 'IMT-Advanced'.

Spectrum Requirement Planning in Wireless Communications   Edited by Hideaki Takagi and Bernhard H. Walke
© 2008 John Wiley & Sons, Ltd

In preparation for the WRC-07 Agenda Item 1.4, ITU-R introduced the RATG approach in Report ITU-R M.2074, which is discussed in this section.

7.1.1   Justification for RAT group approach

A key requirement in the development of the spectrum calculation methodology is that the methodology is technology neutral and generic. This means that the calculations should focus on 'future development of IMT-2000' and on 'IMT-Advanced' without addressing individual technologies separately. Another key requirement for the methodology is to have the 'flexibility to handle both emerging technologies, and well characterized systems such as those defined in Recommendation ITU-R M.1457'. This requirement also suggests that the technical characteristics should be maintained on a generic level.

Furthermore, the requirements set for the methodology imply that the methodology should:

- produce results in a manner that is easily understandable and credible;

- be implementable and verifiable within the available time scales;

- be suitable to be used during meetings of ITU-R WP 8F;

- be no more complex than is justified by uncertainty of the input data.

The first conclusion drawn from the requirements set for the methodology is that a suitable, technology neutral and simplified approach is to have the different radio technologies grouped instead of considering possibly tens of technologies individually.

The second conclusion is that in order to cover the scope of the WRC-07 Agenda Item 1.4, there is a need for the following RATGs: 'Future development of IMT-2000' and 'IMT-Advanced'.

According to Resolution 228 (Rev. WRC-03), the service functionalities in fixed, mobile and broadcasting networks are increasingly converging and interworking. This implies a need to take also other RATGs into account. The fixed, mobile and broadcasting networks can increasingly cover overlapping service types and therefore, the services and traffic of other relevant converging technologies must be considered.

However, the WRC-07 Agenda Item 1.4 concentrates on the future development of IMT-2000 and IMT-Advanced and the aim is therefore not to calculate the spectrum requirements of the other technologies. The third conclusion to be drawn is, therefore, that the traffic distribution to the other relevant RATGs must be taken into account. To do so, the other relevant RATGs need to be defined. The definition of RATGs is considered in the next section.

Finally, the RAT grouping approach will ease the process of acquiring the required input data for the spectrum calculation methodology. The methodology requires market data as input to the calculations as explained in Chapter 6. It can be assumed that the predicted traffic can be distributed with reasonable accuracy to relevant RATGs but not realistically down to individual RATs, irrespective of how the distribution is done. Moreover, the methodology requires also radio-related parameters as input to the calculations. Defining the radio-related parameters for a few RATGs is expected to be much more feasible than for tens of separate RATs.

7.1.2 Definition of RAT groups

All the relevant radio technologies need to be covered in the spectrum calculation methodology as discussed in Section 7.1.1. Based on the WRC-07 Agenda Item 1.4, Resolution 228 (Rev. WRC-03), and Recommendation ITU-R M.1645, the following RATGs are identified in Report ITU-R M.2074:

- RATG 1: Pre-IMT systems, IMT-2000 and its future development.

  This group covers the cellular mobile systems, IMT-2000 systems and their enhancements.

- RATG 2: IMT-Advanced.

  This group covers new capabilities of IMT-Advanced, as defined in Rec. ITU-R M.1645, e.g. new mobile access and new nomadic/local area wireless access.

- RATG 3: Existing radio local area networks (RLANs) and their enhancements.

- RATG 4: Digital mobile broadcasting systems and their enhancements.

  This group covers systems aimed at broadcasting to mobile and handheld terminals.

The justifications for the establishment of the presented four RATGs are the following:

RATG 1: The need for RATG 1 stems directly from the WRC-07 Agenda Item 1.4 and Rec. ITU-R M.1645. The inclusion of IMT-2000 and its future development into a single RATG is in line with the expectation of Rec. ITU-R M.1645 stating that 'There will be a steady and continuous evolution of IMT-2000 to support new applications, products and services'. This evolution is also confirmed by ongoing standardization activities.

RATG 2: The need for RATG 2 also stems directly from the WRC-07 Agenda Item 1.4 and Rec. ITU-R M.1645. The motivation for introducing a separate RATG for IMT-Advanced in addition to RATG 1 is that IMT-Advanced is expected to have significantly differing RAT characteristics and capabilities than those of IMT-2000 and its future development.

RATG 3. The need for taking RATG 3 into account comes from Rec. ITU-R M.1645. One of the new capabilities of IMT-Advanced is the new nomadic/local area wireless access. It can be expected that existing nomadic/local area wireless access systems will share a portion of the relevant traffic. The WRC-03 identified globally common spectrum for them, which allows considerable capacity for such networks.

RATG 4: The need for taking RATG 4 into account comes from Rec. ITU-R M.1645 and the emergence of new mobile broadcasting services. The new mobile broadcasting services providing point-to-multipoint services based on technologies such as IP datacast are expected to emerge in the coming years forming a part of the total mobile market, which justifies the need for this RATG.

Pre-IMT technologies are not included as a separate RATG, but they are incorporated into RATG 1 together with IMT-2000 and its future development. The reasons are as follows:

- Pre-IMT systems cover a subset of the IMT-2000 services and therefore the corresponding traffic can be aggregated with IMT-2000 traffic.

- Most bands for pre-IMT-2000 technologies are identified for IMT-2000, and as such those bands will be taken into account in the spectrum estimations anyway.

- The presence of pre-IMT systems can technically be taken into account by appropriate adjustments in radio parameters of RATG 1, e.g. the spectral efficiency, so that the value of each radio parameter is representative for all RATs in the group.

- The time span for the market data is beyond 2015 when the significance of the pre-IMT systems may be decreasing in some countries or Regions. However, there will be differences in different countries and Regions with respect to the licensing, market development, migration to IMT-2000, etc. Covering such questions was not in the scope of the WRC-07 Agenda Item 1.4.

7.1.3   Usage of RAT groups

All of the four RATGs described in the preceding section are relevant for the spectrum calculation process and necessary up to the traffic distribution part, i.e. Step 4 of the methodology flow chart in Figure 4.2. RATG 1 and RATG 2 will be covered through the whole calculation process, and the result of the spectrum requirement calculation will cover the spectrum requirements for pre-IMT, IMT-2000, future development of IMT-2000 and IMT-Advanced in accordance with the WRC-07 Agenda Item 1.4.

The traffic distribution to RATGs is done based on the service and market predictions, within the methodology or by both together. Traffic distribution based on the market data would be the preferred choice, but it is very unlikely that all required information would be available from the market predictions. That was indeed eventually the case with Report ITU-R M.2072.

The traffic distribution to RATGs needs to consider the availability and capabilities of the different RATGs in the different service environments. Market predictions can provide a first indication of which service categories each RATG can support. The final distribution of traffic to RATGs is derived by considering both market data and radio aspects.

7.2   Use of Radio Parameters in the Methodology

The ITU spectrum calculation methodology uses a limited set of radio parameters to characterize the existing, emerging and future wireless systems. The radio parameters are assumed to be the same for uplink and downlink directions. RATG 1 and RATG 2 are characterized with more radio parameters than RATG 3 and RATG 4 since the spectrum requirements are calculated only for RATG 1 and RATG 2. The spectrum requirements of RATG 3 and RATG 4 are out of the scope of the presented methodology. In the following, the meaning and role of the different radio parameters used in the methodology is explained.

7.2.1 Cell area

The cell area parameter presents the area covered by the macro cell, micro cell, pico cell and hot spot radio environments (REs) in km^2/cell. The deployment of the radio environments depends on the teledensity, and therefore the values of the cell area parameter are given separately for the different teledensities. The cell area parameter is independent of the RATGs considered. The cell area parameter is used to calculate the offered traffic load in the different radio environments of the different RATGs in different teledensities, based on the traffic density figures from market studies. The session arrival rate per cell in sessions/s/cell for unicast traffic is obtained as the product of the traffic distribution ratio, user density, session arrival rate per user and cell area. Therefore, the cell area has a direct influence on the traffic per cell according to Equation (4.13). When the cell area is increased, the session arrival rate per cell increases, and thus also the traffic volume per cell increases leading to an increase in the unadjusted spectrum requirements.

The spectrum calculations are sensitive to the cell area parameter and therefore realistic values for the cell area parameter should be considered. The cell area is one of the most influential input parameters and therefore it is included in the front sheet 'Main' in the spectrum calculation MS Excel tool as explained in Section 5.2.

Realistic values for the cell area parameter should be obtained with link budget calculations. Values for the cell area parameter should take into account the operating environment, Quality of Service (QoS) criteria, and system characteristics. The operating environment should be characterized with propagation conditions in the given deployment and interference situations. The QoS criteria should consider target data rates such as the peak data rate and the cell edge user data rate to guarantee certain minimum performance for all users. System characteristics should consider antenna configurations, transmitter and receiver performance, and carrier bandwidths which depend on the data rate. The values for cell area should also take into account the requirements for mobility support in different cell types. For example, the macro cell is defined to support all mobility classes from stationary/pedestrian to high mobility which sets a lower limit on the available cell size for macro cell deployment.

7.2.2 Application data rate

The application data rate is a performance measure for the different cell types of the RATGs. The application data rate represents a bit rate in bit/s which is available for service categories in a particular radio environment in a given RATG. The application data rate parameter is used in the methodology in the distribution of traffic to RATGs and radio environments. Therefore, it is needed for all four RATGs. The application data rate parameter determines whether a service category can be supported by a given RATG in a given radio environment by comparing the requirements of the service categories in terms of the mean service bit rate with the RATG capabilities in terms of the application data rate as explained in Section 4.3.2.

The market studies characterize the service categories with only one type of data rate parameter which is the mean service bit rate. The mean service bit rate of the service category presents an average data rate requirement which is obtained as the weighted average of different services belonging to the same service category. Only one type of data rate parameter is therefore used in the methodology to model the RATG to reduce the complexity. The application data rate may be smaller than the available peak bit rate and may not be available throughout the whole cell. However, the application data rate needs to be sufficiently large to accommodate the service categories from the market studies which can be supported by future systems.

When the application data rates are small compared to the mean service bit rates required by the service categories, the service categories cannot be supported by the given RATGs in the given radio environments. As a result, the traffic of these service categories cannot be distributed to the RATGs, which means that the future services cannot be served by the future systems. Therefore, the application data rates of the future systems should be larger than the mean service bit rate requirements of the users in order to meet the demand of future services.

Changing the values for the application data rate parameter changes the availability of the different radio environments in the different RATGs. By setting the application data rate of a certain radio environment in a certain RATG to be equal to zero, the corresponding radio environment is not available in the given RATG. This can be done to turn off certain radio environments, which in turn changes the traffic distribution functionality, leading to different unadjusted spectrum requirements. The 'RATG1&2Def-Input' worksheet (sheet 5) of the MS Excel tool includes the application data rate parameter values separately for the three different forecast years in order to model the differences in the availability of the RATGs over the years as described in Section 5.3.

In the macro and micro cellular environments, the system is operating under interference-limited conditions. There the available aggregate throughput decreases significantly with increased distance. Therefore the application data rate, especially in the macro cell environment, corresponds to an expected average aggregate throughput, which is smaller than the peak aggregate throughput of the system. In the micro cell environment, potentially higher physical layer modes[1] can be applied compared to macro cells due to higher expected carrier-to-interference ratios, which in general enable higher application data rates than in macro cells.

The pico cell and hot spot correspond to indoor scenarios where the system is operating under noise-limited conditions well above the noise level. Therefore the application data rate corresponds to the required peak aggregate throughput of the system which is available in the entire indoor cell with high probability with respect to the expected short range under realistic assumptions for indoor applications.

7.2.3   Spectral efficiency

The spectral efficiency parameter is another performance measure for the different cells of the RATGs in addition to the application data rate parameter described in Section 7.2.2. The spectral efficiency presented in bit/s/Hz/cell defines how efficiently the RATG can use the available spectrum in terms of how many bits per second can be communicated over a given bandwidth in a cell. In the spectrum calculation methodology, the spectral efficiency parameter is used to calculate the unadjusted spectrum requirements per cell by dividing the

---

[1] Higher physical layer modes mean that we select a sub-carrier modulation technique that includes more bits per symbol; see, e.g., Walke (2002, p. 815). Therefore, one symbol carries more data and the data rate can be increased. This is possible in better channel conditions such as in this case in the micro cell compared to the macro cell.

capacity requirement in bit/s/cell by the spectral efficiency values in bit/s/Hz/cell as shown in Equation (4.39). Therefore, the spectral efficiency is needed for only RATG 1 and RATG 2. The spectral efficiency is determined from the mean data throughput achieved over all users, who are homogeneously distributed in the area of the radio environment, on the Internet Protocol (IP) layer for packet-switched services, and on the application layer for circuit-switched services.

The capacity requirement in Equation (4.39) presents the aggregate capacity requirement of uplink and downlink, and thus the corresponding spectral efficiency is independent of the link direction. The capacity requirement is the average aggregate capacity required in the cell which is calculated from traffic figures that present the average aggregate traffic over the cell. Also, the spectral efficiency is presented in a single figure per cell in an average sense which characterizes the situation over the whole cell. The calculation of the spectral efficiency values is based on the average aggregate throughput of all users in the cell normalized with the bandwidth of the cell.

The spectral efficiency parameter has a big influence on the spectrum requirements. If the spectral efficiency is increased, the unadjusted and final spectrum requirements decrease. A decrease in the spectral efficiency values results in increased spectrum requirements. The actual values for the spectral efficiency parameter should consider the QoS criteria, including sufficient user satisfaction and data rate requirements, the operating environment, including propagation conditions and interference situation, and the system characteristics in terms of antenna configurations and transmitter/receiver performance.

7.2.4 Minimum spectrum deployment per operator per radio environment

The minimum spectrum deployment per operator per radio environment parameter is the minimum amount of spectrum in MHz needed for one operator to build a practical working network in a given radio environment. The minimum deployment per operator per radio environment is a spectrum granularity unit related to the carrier bandwidth. This parameter is used in Step 7 of the methodology flow chart in Figure 4.2, where the adjustments on the spectrum requirements are made. Therefore, this parameter is only used to characterize RATG 1 and RATG 2.

The minimum deployment parameter has a big influence on the final spectrum requirements. In the end, the final spectrum requirements are integer multiples of this parameter. The spectrum requirements are rounded up in Equation (4.42) so that the last carrier which may be only lightly loaded also receives the full amount of spectrum equal to the minimum deployment per operator per radio environment. Choosing a large value for this parameter may therefore lead to significantly larger final spectrum requirements. Lower values of the parameter in general lead to smaller spectrum requirements because the last carrier which is not fully loaded requires less spectrum. However, there are exceptions where an increase in the minimum deployment parameter leads to a decrease in the final spectrum requirements. This is because in the default situation the last carrier may be only lightly loaded while with increased minimum deployment the last carrier becomes heavily loaded but leading to smaller total spectrum requirement because the number of carriers is lower.

The derivation of the values for the minimum deployment parameter needs to ensure that the application data rate can be supported in the given radio environment. In addition, the minimum deployment should consider the cell edge bit rate to ensure reasonable user satisfaction, also for those users who are located at cell edge.

7.2.5   Number of overlapping network deployments

The number of overlapping network deployments represents the number of coexisting networks of the same RATG that do not share the spectrum. This parameter is used in Steps 7 and 8 of the methodology flow chart in Figure 4.2 to adjust the spectrum requirements according to Equations (4.41)–(4.44) in Section 4.3.6. It is needed only for RATG 1 and RATG 2.

The number of overlapping network deployments has an important influence on the final spectrum requirements. The unadjusted spectrum requirement is divided by the number of overlapping network deployments to obtain the unadjusted spectrum requirement per operator. This quantity is rounded up into an integer multiple of the minimum deployment per operator per radio environment parameter. The spectrum requirements of the overlapping network deployments are then aggregated.

If the number of overlapping network deployments is increased, the final spectrum requirements are also increased. The increase in the spectrum requirement can be significant depending on the minimum deployment per operator per radio environment parameter which shows the spectrum granularity per network deployment.

The methodology assumes that the different overlapping network deployments of a given RATG do not share the spectrum. Some form of spectrum sharing may be implemented in future wireless systems. To take this into account, the number of overlapping network deployments in the calculations should be kept low to prevent an overestimation of the spectrum requirements.

7.2.6   Other radio parameters

In addition to the radio parameters described in Sections 7.2.1–7.2.5, the spectrum calculation methodology also defines the maximum supported velocity, support for multicast, and guard band between operators.

Maximum supported velocity

The maximum supported velocity in km/h defines an upper limit to the supported velocity of the mobile terminals in a given radio environment of a given RATG. Maximum supported velocity is used in the traffic distribution part of the methodology. It is needed for all four RATGs. It is used to identify available radio environments that are capable of supporting traffic with given mobilities. Namely, the maximum supported velocity of the macro cell radio environment of the RATG is compared with a threshold to ensure that the high mobility class services are supported in macro cells.

Support for multicast

The support for multicast parameters defines whether a given RATG is capable of providing multicast transmission, i.e. transmitting multicast traffic to multiple users simultaneously. Support for multicast is used in the traffic distribution part of the methodology which is performed separately for unicast and multicast traffic. It is needed for all four RATGs. The parameter is used to identify the RATGs that are available to serve multicast traffic. Multicast traffic is allocated only to RATGs which support multicast transmission. If there is multicast traffic predicted in the market studies and a RATG can support multicast traffic with the given data rate requirements, the multicast traffic is allocated to the RATG, leading to an increase in the spectrum requirements compared to the situation without multicast support.

Guard band between operators

The guard band between operators defines the excess bandwidth that must be left between the operating bands of two operators in order not to cause harmful interference. The guard band between operators is used in Step 8 of the methodology flow chart in Figure 4.2 to adjust the spectrum requirements in Equation (4.44). Therefore, it is needed for only RATG 1 and RATG 2. If the number of operators is equal to one, the guard band between operators does not influence the calculations. A non-zero guard band between operators directly increases the spectrum requirements if the number of operators exceeds one.

7.2.7 Relations of radio parameters

The radio parameters of the different RATGs are used to model the real wireless networks in a simplified manner. The radio parameters, their units and dimensions and use in the methodology are summarized in Figure 7.1. The radio parameters are closely interrelated. The derivation of the values for these input parameters in order to be used in calculating the actual spectrum requirements should therefore be conducted within the same framework. The input parameter values should take into account the operating environment including the propagation conditions and interference situation, QoS criteria including data rate requirements and user satisfaction, and system characteristics in terms of antenna configurations and transmitter/receiver performance. It is important that the same situation is considered when deriving the values for the different input parameters due to the interrelations. For example, higher data rates can be achieved at the cost of reducing the cell size.

The radio parameters have a large influence on the spectrum requirements. The application data rate parameter determines the availability of RATGs and radio environments for different service categories. The cell area parameter influences the traffic volumes significantly.

The spectral efficiency has a significant impact as the capacity requirement is divided by the spectral efficiency to determine unadjusted spectrum requirements. The minimum deployment per operator per radio environment adjusts the spectrum requirements and presents the spectrum granularity. In the end, the final spectrum requirements are integer multiples of this parameter. The number of overlapping network deployments gives the number of networks that do not share the spectrum and thus require their own spectrum, which is at least equal to the minimum deployment per operator per radio environment. If the number of overlapping network deployments exceeds one, the guard band between operators also influences the final spectrum requirements.

Radio parameter	Required RATG	Unit	Dimension	Use in methodology	
Cell area	Independent of RATGs	km²/cell	RE, teledensity	Equation (4.13)	Traffic volume calculation
Application data rate	All RATGs	bit/s	RATG, RE, forecast year	Section 4.3.2	
Maximum supported velocity	All RATGs	km/h	RATG, RE	Section 4.3.2	Traffic distribution
Support for multicast	All RATGs	yes/no	RATG	Section 4.3.2	
Spectral efficiency	RATG 1 RATG 2	bit/s/Hz/cell	RATG, RE, teledensity, forecast year	Equation (4.39)	
Number of overlapping network deployments	RATG 1 RATG 2	—	RATG	Equation (4.41) Equation (4.44)	Spectrum requirement calculation
Minimum spectrum deployment	RATG 1 RATG 2	MHz	RATG, RE	Equation (4.42)	
Guard band between operators	RATG 1 RATG 2	MHz	RATG	Equation (4.44)	

Figure 7.1 Radio parameters used in ITU spectrum calculation methodology.

7.3   Example Input Radio Parameter Value Set

The ITU parameter set provided in Report ITU-R M.2078 includes specific parameter values for the calculations. In the following, an example set of the parameter values for the radio-related input parameters is presented. The values are given for the radio parameters listed in Section 7.2 as well as for two other parameters, namely the population coverage percentage and distribution ratio among available RATGs, which are closely related to the radio parameters. The parameter values are further used in the numerical example calculations in Chapter 8.

7.3.1   Radio parameters

The cell areas of the different radio environments in the different teledensities expressed in km²/cell are given in Table 7.1. The cell areas are input for the front sheet 'Main' in the MS Excel tool. The example cell area takes into account the *penetration loss* (that also includes the path loss) which reduces the coverage distance. The penetration losses of 18, 15 and 12 dB are used for dense urban, suburban and rural teledensities, respectively. The cell areas of pico

Table 7.1   Cell area of radio environments in km^2/cell according to the penetration loss in different teledensities.

Radio environment	Dense urban	Suburban	Rural
Macro cell	0.10	0.15	0.22
Micro cell	0.07	0.10	0.15
Pico cell	0.0016	0.0016	0.0016
Hot spot	0.000065	0.000065	0.000065

Table 7.2   Radio parameters for RATG 1.

Radio parameters	Macro cell	Micro cell	Pico cell	Hot spot
Application data rate (Mbit/s)	20	40	40	—
Maximum supported velocity (km/h)	250	50	4	—
Guard band between operators (MHz)	0	0	0	—
Minimum deployment per operator per radio environment (MHz)	40	40	40	—
Support for multicast	yes	yes	yes	—
Number of overlapping network deployments	1	1	1	—

Table 7.3   Radio parameters for RATG 2.

Radio parameters	Macro cell	Micro cell	Pico cell	Hot spot
Application data rate (Mbit/s)	50	100	1000	1000
Maximum supported velocity (km/h)	250	50	4	4
Guard band between operators (MHz)	0	0	0	0
Minimum deployment per operator per radio environment (MHz)	20	20	120	120
Support for multicast	yes	yes	yes	yes
Number of overlapping network deployments	1	1	1	1

cell and hot spot do not depend on the teledensity. This example is taken from Table 15b in Report ITU-R M.2078 (errors corrected).

Tables 7.2, 7.3, 7.4 and 7.5 give the radio parameters of RATG 1, RATG 2, RATG 3 and RATG 4, respectively. These tables are taken from Tables 18, 19, 20 and 21, respectively, in Report ITU-R M.2078. Tables 7.2 and 7.3 are input for the 'RATG1&2Def-Input' worksheet, while Tables 7.4 and 7.5 are input for the 'RATG3&4Def-Input' worksheet in the MS Excel tool. The parameter number of the overlapping network deployment is input for the front sheet 'Main'. The same RATG parameter values are used for the three forecast years. However, when a certain RATG is turned off in a particular forecast year, the application data rate is set to zero for the corresponding year in the 'RATG1&2Def-Input' or 'RATG3&4Def-Input' worksheet in the MS Excel tool.

Table 7.4   Radio parameters for RATG 3.

Radio parameters	Macro cell	Micro cell	Pico cell	Hot spot
Application data rate (Mbit/s)	—	—	50	100
Maximum supported velocity (km/h)	—	—	4	4
Support for multicast	—	—	yes	yes

Table 7.5   Radio parameters for RATG 4.

Radio parameters	Macro cell
Application data rate (Mbit/s)	2
Maximum supported velocity (km/h)	250
Support for multicast	yes

RATG 1 does not include parameter values for the hot spot radio environment because it is assumed that the hot spot does not exist in RATG 1. Moreover, RATG 2 is assumed to be unavailable in 2010. Therefore its application data rate is set to zero in the 'RATG1&2Def-Input' worksheet in all radio environments in the corresponding year.

The spectral efficiency values for RATG 1 in unicast and multicast transmission modes for the years 2010, 2015 and 2020 are given in Table 7.6. The spectral efficiency values for RATG 2 in unicast and multicast transmission modes for the years 2010, 2015 and 2020 are given in Table 7.7. These tables are taken from Tables 22 and 23, respectively, in Report ITU-R M.2078. The spectral efficiency values are input for 'RATGEff-Input' worksheet in the MS Excel tool. In general, the spectral efficiency values can be different in different teledensities, but the example parameter value set assumes the same values in different teledensities. The spectral efficiencies of RATG 1 and RATG 2 are assumed to improve over time as the systems are developed further. The multicast spectral efficiencies are assumed to be half of the unicast spectral efficiencies because in multicasting the radio resources are assumed to be used according to the user with the lowest reception condition.

Table 7.6   Spectral efficiency of RATG 1 in bit/s/Hz/cell.

Years	Modes	Macro cell	Micro cell	Pico cell	Hot spot
2010	Unicast	1	2	2	—
	Multicast	0.5	1	1	—
2015	Unicast	1.5	3	3	—
	Multicast	0.75	1.5	1.5	—
2020	Unicast	2	4	4	—
	Multicast	1	2	2	—

Table 7.7   Spectral efficiency of RATG 2 in bit/s/Hz/cell.

Years	Modes	Macro cell	Micro cell	Pico cell	Hot spot
2010	Unicast	2	2.5	3	5
	Multicast	1	1.25	1.5	2.5
2015	Unicast	4.25	5.5	7	8.25
	Multicast	2.125	2.75	3.5	4.125
2020	Unicast	4.5	6	7.5	9
	Multicast	2.25	3	3.75	4.5

Table 7.8   Population coverage percentage of radio environments in different service environments in 2010.

Service environments	Macro cell	Micro cell	Pico cell	Hot spot
SE 1	100	90	0	80
SE 2	100	90	20	80
SE 3	100	95	20	10
SE 4	100	15	0	80
SE 5	100	40	35	20
SE 6	100	0	10	50

Table 7.9   Population coverage percentage of radio environments in different service environments in 2015.

Service environments	Macro cell	Micro cell	Pico cell	Hot spot
SE 1	100	90	10	80
SE 2	100	90	20	80
SE 3	100	95	30	25
SE 4	100	35	0	80
SE 5	100	50	35	20
SE 6	100	0	10	50

7.3.2   Population coverage percentage and traffic distribution ratio

The population coverage percentages of the different radio environments in different service environments are given in Tables 7.8, 7.9 and 7.10 for the years 2010, 2015 and 2020, respectively. These are taken from Table 16 of Report ITU-R M.2078. The population coverage percentages are input into 'SE-Input' worksheet in the MS Excel tool. The value '0' denotes that the given combination of radio environment and service environment does not exist.

The population coverage percentages are assumed to evolve over time as the deployment of the new systems develops. Certain combinations, such as a pico cell radio environment in a dense urban home (SE 1), are assumed to become available only later in time. Other combinations, such as a pico cell in a dense urban public area (SE 3), are assumed to obtain increased coverage over time.

Table 7.10   Population coverage percentage of radio environments in different service environments in 2020.

Service environments	Macro cell	Micro cell	Pico cell	Hot spot
SE 1	100	90	20	80
SE 2	100	90	20	80
SE 3	100	95	40	40
SE 4	100	35	0	80
SE 5	100	50	35	20
SE 6	100	0	10	50

Table 7.11   Distribution ratios among available RATGs in 2010.

Available RATGs	RATG 1	RATG 2	RATG 3
1	100	—	—
2	—	100	—
3	—	—	100
1, 2	100	0	—
1, 3	30	—	70
2, 3	—	0	100
1, 2, 3	30	0	70

Table 7.12   Distribution ratios among available RATGs in 2015.

Available RATGs	RATG 1	RATG 2	RATG 3
1	100	—	—
2	—	100	—
3	—	—	100
1, 2	50	50	—
1, 3	20	—	80
2, 3	—	30	70
1, 2, 3	20	20	60

The traffic distribution ratios among the available RATGs are given in Tables 7.11, 7.12 and 7.13 for the years 2010, 2015 and 2020, respectively. These tables are taken from Tables 24a, 24b and 24c, respectively, in Report ITU-R M.2078. The distribution ratios among the available RATGs are input into the 'RATG-DistRatio-Input' worksheet in the MS Excel tool. This parameter cannot be used to turn off certain RATGs in certain forecast years. The exclusion of RATG 2 in 2010 in the ITU parameter set is done with the application data rate parameter as explained in Section 7.2.2.

The traffic distribution ratios among available RATGs reflect the different market shares of the RAT groups. The traffic distribution ratios among available RATGs are assumed to

Table 7.13   Distribution ratios among available RATGs in 2020.

Available RATGs	RATG 1	RATG 2	RATG 3
1	100	—	—
2	—	100	—
3	—	—	100
1, 2	10	90	—
1, 3	10	—	90
2, 3	—	50	50
1, 2, 3	10	45	45

evolve over time. RATG 2 is assumed to be available only from 2015 onwards. Part of the traffic delivered by RATG 1 is assumed to shift to RATG 2 and RATG 3 over time, as well as from RATG 3 to RATG 2.

8

Numerical Examples

Tim Irnich, Marja Matinmikko, Jussi Ojala and Bernhard H. Walke

In this chapter the spectrum requirement calculation methodology is illustrated by means of an example calculation using the calculation tool described in Chapter 5 and the scenarios described in Chapters 6 and 7 which include the example sets of market and radio-related input parameters. For the sake of brevity the spectrum requirement calculation for the higher user density setting for the year 2020 from Report ITU-R M.2078 is taken as an example. Where appropriate, the illustration of the calculation process is limited to the downlink transmission direction of service environment (SE) 1 (i.e. up to Equation (4.15) in Section 4.3.3) and the dense urban teledensity (i.e. from Equation (4.16) onwards in Section 4.3.3), respectively. Since the algorithm discussed here is designed to calculate the spectrum requirements for Radio Access Technique Group (RATG) 1 and RATG 2 only, the presentation is limited to RATG 1 and RATG 2 from the calculation of the required system capacity onwards.

Section 8.1 complements the scenario used for estimation of the spectrum requirements for the future development of IMT-2000 and IMT-Advanced by introducing the example values for the remaining input parameters in addition to the market parameters and the radio-related parameters introduced in Chapters 6 and 7, respectively. We present the maximum allowable blocking probability for circuit-switched service categories (SCs) 1–10, and the IP packet size statistics along with the maximum allowable mean packet delay for the packet-switched SCs 11–20. In the subsequent sections, the intermediate and final results of the calculation algorithm are described step by step. Section 8.2 shows the parameters that describe the offered traffic, including mobility ratios, derived from the ITU market study for the year 2020 in SE 1, given in Report ITU-R M.2078. Section 8.3 illustrates the calculation of traffic distribution ratios for SE 1. Section 8.4 shows the resulting offered traffic in RATG 1 and RATG 2. Section 8.5 explains the calculation of required system capacity and its results.

Spectrum Requirement Planning in Wireless Communications   Edited by Hideaki Takagi and Bernhard H. Walke
© 2008 John Wiley & Sons, Ltd

Finally, Section 8.6 presents the resulting spectrum that is needed to provide the required system capacity, taking into account practical network deployments.

8.1 Packet Size Statistics and QoS Requirements

In addition to the market-related service category parameters described in Chapter 6, the capacity calculation algorithm of the methodology also requires other service category parameters which cannot be obtained from Report ITU-R M.2072 as described in Section 4.2.1. Different parameters are needed for circuit-switched and packet-switched service categories.

For service categories SC 1–10 treated as circuit-switched, the only additionally required parameter is the maximum allowable blocking probability. The assumed value is 0.01 for both conversational and streaming traffic classes as shown in Table 8.1. For packet-switched service categories SC 11–20, the additionally required parameter values are the mean and second moment of the packet size distribution and the maximum allowable mean packet delay. They are given in Table 8.2 for the interactive traffic class (SC 11–15) and in Table 8.3 for the background traffic class (SC 16–20).

Table 8.1   Parameters for conversational and streaming traffic classes (SC 1–10).

Traffic class	Service categories	Maximum allowable blocking probability
Conversational	SC 1–5	0.01
Streaming	SC 6–10	0.01

Table 8.2   Parameters for interactive traffic class (SC 11–15).

Service type	Mean packet size	Second moment of packet size	Maximum allowable mean packet delay
Super-high multimedia	3292.23	27 552 481.16	0.1490
High multimedia	1847.82	15 349 865.20	0.1019
Medium multimedia	1021.60	6 592 429.07	1.5280
Low rate data/low multimedia	102.56	138 595.74	2.7813
Very low rate data	47.61	36 019.39	0.4224

The values for the IP packet size statistics are derived as follows. Analyzing the representative applications identified in the market study (see Chapter 6), it is obvious that many representative applications are very similar in terms of underlying technology, traffic characteristics and Quality of Service (QoS) requirements. Thus, it is possible to identify a rather small number of real-world example applications so that either one or a mixture of two of them (e.g. video streaming mixed with audio streaming represents a mobile TV

Table 8.3   Parameters for background traffic class (SC 16–20).

Service type	Mean packet size	Second moment of packet size	Maximum allowable mean packet delay
Super-high multimedia	3054.00	20 332 660.50	0.0648
High multimedia	3307.86	27 691 445.33	0.4968
Medium multimedia	1369.33	11 523 733.33	2.9670
Low rate data/low multimedia	235.50	1 827 768.50	4.9444
Very low rate data	235.50	1 827 768.50	44.5000

representative application) represents a certain subset of the representative applications in terms of IP packet size distribution. By studying the traffic characteristics of these real-world examples, we can determine the mean and the second moment of the IP packet size distribution. The mean delay requirements are calculated by considering the mean packet size of the real-world examples and the mean service bit rate of the representative application.

8.2   Traffic Demand Derived from Market Data

The example calculation uses the market study parameter value set based on the higher user density scenario in the year 2020 from Report ITU-R M.2078. The market study parameter value set is introduced in Section 6.4 and the actual values for the parameters are given in Appendix B. In the following sections the five different market study parameters are illustrated. For simplicity, the treatment of the traffic demand from market data is restricted to the unicast traffic in the downlink direction in SE 1 (dense urban home). The corresponding numerical values are given in Tables B.1–B.3 in Appendix B.

8.2.1   User density

The market setting for the user density parameter U presented in Section 6.4 (Table 6.2) is equal to 25% in all service categories, meaning that the actual values for the user density parameter are taken from the lower part of the ranges given in the market study. The values of the user density are shown in Figure 8.1 for all service categories in SE 1. Although not visible in the figure owing to the large range of values, zero user density only occurs for SC 1 and SC 16. The highest user density is assumed for SC 5 and SC 12.

8.2.2   Session arrival rate per user

The market setting for the session arrival rate per user Q presented in Section 6.4 (Table 6.2) is equal to 30% in all service categories. Figure 8.2 shows the values for the session arrival rate per user for all service categories in SE 1. The most often used service categories are SC 12 and SC 15. Since SC 12 also has the second highest user density (see Figure 8.1), SC 12 is likely to offer significantly more traffic than any other SCs, but of course this also depends on the other traffic parameters such as the average session duration and mean service bit rate.

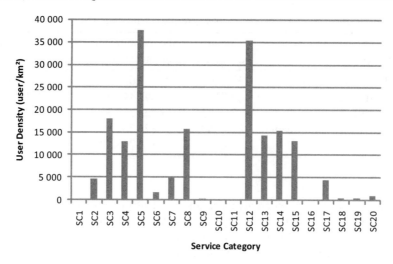

Figure 8.1 User density per SC in SE 1 in the year 2020.

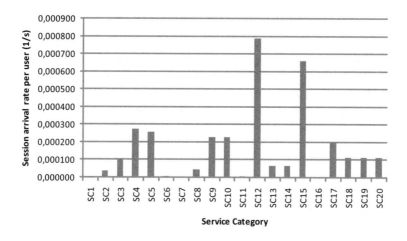

Figure 8.2 Session arrival rate per user per SC in SE 1 in the year 2020.

8.2.3   Average session duration

The market setting for the average session duration parameter μ presented in Section 6.4 (Table 6.2) is equal to 30% in all service categories. Figure 8.3 shows the values for the average session duration for all service categories in SE 1. The longest session durations are assumed for SC 4 and SC 7. Many of the packet-switched service categories (SC 11–20) assume rather low average session durations in the order of 20 to 30 s/session with only a few exceptions. The session durations of the circuit-switched service categories (SC 1–10) are considerably longer.

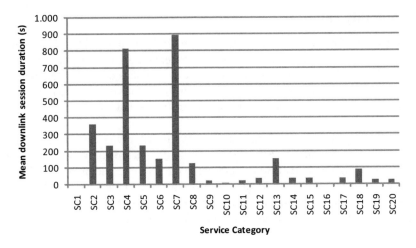

Figure 8.3 Average session duration per SC in SE 1 in the year 2020.

8.2.4 Mean service bit rate

The market setting for the mean service bit rate parameter r presented in Section 6.4 (Table 6.2) is equal to 30% in all service categories. Figure 8.4 shows the values for the mean service bit rate for all service categories in SE 1. The highest mean service bit rates are in the order of 300 Mbit/s in SC 6 and SC 11. The mean service bit rates of other service categories vary approximately between 10 kbit/s and 10 Mbit/s. SC 1 and SC 16 do not include any traffic in SE 1 in the year 2020 and therefore the corresponding mean service bit rates are equal to zero.

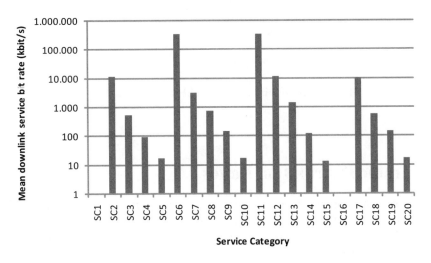

Figure 8.4 Mean service bit rate per SC in SE 1 in the year 2020.

8.2.5 Mobility ratios

The market setting for the mobility ratios presented in Section 6.4 (Table 6.2) assumes the middle mobility scenario for all service categories except for SC 11 from the three alternative lowest, middle and highest mobility scenarios. The exception is due to the fact that SC 11 is a very high data rate service category with more than 100 Mbit/s data rate requirements. Very high data rate services are not supported at higher mobilities but require stationary usage for operation.

Figure 8.5 shows the values for the mobility ratios for all service categories in SE 1. Some of the service categories, such as SC 9–10 and SC 18–20, include only stationary/pedestrian traffic. Moreover, high mobility traffic is present in only SC 4–5 and in SC 13–15 with quite low percentages. The domination of the stationary/pedestrian and low mobility traffic is a special situation for dense urban teledensity areas where also SE 1 belongs. For less densely populated service environments there is a clear shift towards faster movement.

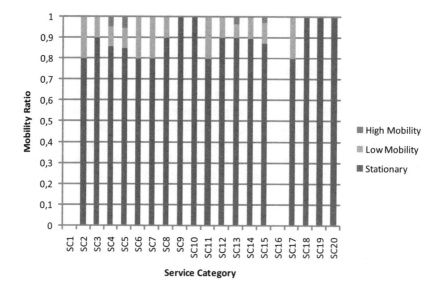

Figure 8.5 Mobility ratios for SE 1 in the year 2020.

8.3 Traffic Distribution Ratios

The possible combinations of service categories, radio environments and RATGs for the example calculation of the unicast traffic in the downlink direction of SE 1 in the year 2020 are shown in Tables 8.4 and 8.5. The combinations are denoted by flags '0' and '1' showing which service categories can be supported by which combination of RATG and radio environment in SE 1, as explained in Sections 4.3.2 theoretically and in Section 5.4 in the Worksheets 17–19 of the calculation tool. The marking '0' denotes that the corresponding

RATG and radio environment combination is not available due, for example, to zero population coverage percentage or too small an application data rate which does not support the required mean service bit rate of the service category. The marking '1' means that the corresponding RATG and radio environment combination is available for the given service category in SE 1.

Table 8.4 Possible combinations of service categories and radio environments in the downlink direction for RATG 1 and RATG 2 in SE 1 in the year 2020.

Service category	RATG 1				RATG 2			
	Macro	Micro	Pico	Hot spot	Macro	Micro	Pico	Hot spot
SC 1–5	1	1	1	0	1	1	1	1
SC 6	0	0	0	0	0	0	1	1
SC 7–10	1	1	1	0	1	1	1	1
SC 11	0	0	0	0	0	0	1	1
SC 12–20	1	1	1	0	1	1	1	1

Table 8.5 Possible combinations of service categories and radio environments in the downlink direction for RATG 3 and RATG 4 in SE 1 in the year 2020.

Service category	RATG 3				RATG 4			
	Macro	Micro	Pico	Hot spot	Macro	Micro	Pico	Hot spot
SC 1–5	0	0	1	1	0	0	0	0
SC 6	0	0	0	0	0	0	0	0
SC 7–10	0	0	1	1	0	0	0	0
SC 11	0	0	0	0	0	0	0	0
SC 12–20	0	0	1	1	0	0	0	0

According to the possible combinations shown in Table 8.4, RATG 1 supports all service categories except for SC 6 and SC 11 due to the very high mean service bit rate requirements (i.e. 321 Mbit/s) of these two service categories. The hot spot radio environment is not assumed to be deployed in RATG 1 and therefore the possible combinations in the hot spot are equal to 0. RATG 2 can support all service categories in all radio environments, with the exception of macro cells and micro cells, where SC 6 and SC 11 are not supported. For RATG 3 in Table 8.5, only pico cells and hot spots are assumed to be deployed. RATG 3 does not support SC 6 and SC 11 owing to the very high mean service bit rate requirements of these service categories compared to the capabilities of RATG 3. RATG 4 is assumed to be a pure broadcasting technology and is therefore treated in a special way in the traffic distribution as such the corresponding table includes values '0' only. The table for RATG 4 is only provided for completeness.

The final traffic distribution ratios are shown in Figure 8.6. The final traffic distribution ratios divide the total traffic among the four radio environments and among RATG 1, RATG 2 and RATG 3. The final distribution ratios have been calculated by the possible combinations

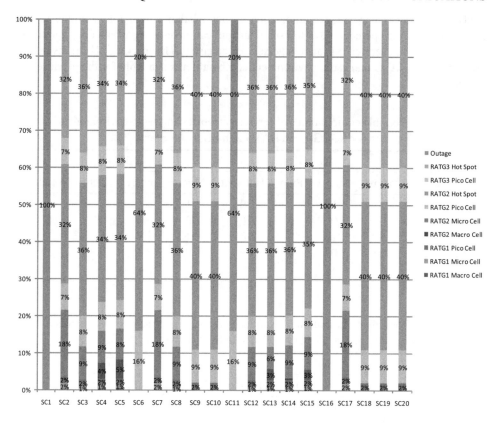

Figure 8.6 Traffic distribution ratios in SE1 in the year 2020.

from Tables 8.4 and 8.5, the mobility ratios from Section 6.4, as well as the population coverage percentages and the traffic distribution ratios among available RATGs defined in Section 7.3.2 by using the approach described in Section 4.3.2.

The traffic in SC 6 and SC 11 is only distributed to pico cells and hot spots of RATG 2. As a general remark from the final traffic distribution ratios, most of the traffic in other service categories is in the hot spot radio environment of RATG 2 and RATG 3 which is a special situation for dense urban teledensity areas. For less densely populated service environments there is a shift towards larger cells. The traffic amounts distributed to RATG 1 are rather low due to the predefined distribution ratios among available RATGs which, in the example calculation, assume that the deployment of RATG 1 is fading in 2020 (see Table 7.13).

8.4 Offered Traffic per RAT Group and Radio Environment

The amount of traffic each RATG is assumed to serve is called the offered traffic and it is calculated for each pair of teledensity and radio environment. This is the needed input to capacity calculations. Since SCs 1–10 are assumed to have more conventional mobile nature than SCs 11–20 which have more Internet type characters, different capacity calculation methods are used for those two groups. Thus the offered traffic needs to be defined differently for those two groups according to the needs of different capacity calculation methods.

For the circuit-switched service categories SC 1–10, the needed input is the average offered traffic per cell (Erlang/cell) and the mean service bit rate (bit/s). For the packet-switched service categories SC 11–20, the aggregated offered traffic per cell (bit/s/cell) is needed. These values are needed for each pair of teledensity and radio environment for RATG 1 and 2 in uplink and downlink directions including unicast and multicast services. In the present example we only concentrate on the downlink unicast services in the dense urban teledensity.

The calculation of the offered traffic is shown in Section 4.3.3. The session arrival rate per cell $P_{m,n,rat,p}$ (sessions/s/cell) in each service environment can be calculated using the input parameters such as the user density, session arrival rate per user, and cell area with distribution ratios according to Equation (4.13) for unicast traffic and to Equation (4.15) for multicast traffic. For the circuit-switched service categories, the average traffic per cell $\rho_{d,n,rat,p}$ (Erlang/cell) in each teledensity for each service category can be calculated according to Equation (4.16) by taking into account the average session duration. The aggregate mean service bit rate $r_{d,n,rat,p}$ (bit/s) in each teledensity for each service category is the weighted average of service bit rates of service environments belonging to the same teledensity as shown in Equation (4.17). For the packet-switched service categories, the aggregated offered traffic $T_{d,n,rat,p}$ (bit/s/cell) is calculated by Equation (4.18).

Table 8.6   Offered circuit-switched traffic per cell in the dense urban teledensity in the year 2020 (Erlang/cell).

Service category	RATG 1			RATG 2			
	Macro	Micro	Pico	Macro	Micro	Pico	Hot spot
SC 1	0.00	0.00	0.00	0.00	0.00	0.00	0.00
SC 2	0.00	0.90	0.06	0.00	8.13	0.29	0.05
SC 3	0.00	1.23	0.07	0.00	11.09	0.33	0.05
SC 4	4.67	7.54	0.43	42.07	67.85	1.92	0.19
SC 5	6.23	7.44	0.43	56.09	67.00	1.91	0.22
SC 6	0.00	0.00	0.00	0.00	0.00	0.00	0.00
SC 7	0.71	1.23	0.09	6.38	11.08	0.39	0.07
SC 8	3.28	3.24	0.20	29.56	29.17	0.91	0.08
SC 9	0.00	0.00	0.00	0.00	0.01	0.00	0.00
SC 10	0.00	0.00	0.00	0.00	0.00	0.00	0.00

Table 8.6 shows the offered traffic in Erlang/cell of circuit-switched service categories SC 1–10 in the dense urban teledensity in different radio environments. The RATG 2 has clearly more traffic mainly owing to the larger distribution ratios. The offered traffic is dominated by SC 4 and SC 5 followed by SC 8.

Tables 8.7 and 8.8 show the offered traffic in kbit/s/cell of packet-switched service categories SC 11–20 in the dense urban teledensity in different radio environments. Clearly RATG 2 has the major part of offered traffic, e.g. in micro cells which dominate the offered

Table 8.7   Offered packet-switched traffic per cell for RATG 1 in the dense urban teledensity in the year 2020 (kbit/s/cell).

Service category	RATG 1		
	Macro cell	Micro cell	Pico cell
SC 11	0.00	0.00	0.00
SC 12	0.00	93 956.46	5270.37
SC 13	375.36	636.89	38.85
SC 14	12.95	171.04	8.27
SC 15	7.41	29.51	1.26
SC 16	0.00	0.00	1.69
SC 17	33.61	4161.04	296.64
SC 18	0.00	0.08	4.57
SC 19	0.01	0.02	0.04
SC 20	0.00	0.01	0.01

Table 8.8   Offered packet-switched traffic per cell for RATG 2 in the dense urban teledensity in the year 2020 (kbit/s/cell).

Service category	RATG 2			
	Macro cell	Micro cell	Pico cell	Hot spot
SC 11	0.00	11 834.92	573.58	16.28
SC 12	0.00	845 608.18	23 716.68	2545.94
SC 13	3378.20	5731.97	174.81	18.52
SC 14	116.51	1539.34	37.20	1.82
SC 15	66.69	265.62	5.65	0.54
SC 16	0.00	0.00	7.60	1.37
SC 17	302.51	37 449.35	1334.88	214.80
SC 18	0.00	0.69	20.55	3.71
SC 19	0.08	0.17	0.18	0.03
SC 20	0.00	0.06	0.04	0.01

traffic in dense urban teledensity. RATG 2 has packet-switched traffic more than nine times higher than the packet-switched traffic of RATG 1. In the micro cell environment the SC 12 clearly offers more traffic than any other packet-switched service category.

8.5   Required System Capacity

After the amount of offered traffic is determined for each combination of RATG and radio environment, the amount of system capacity that is needed to provide the required QoS is determined. The calculation is done separately for circuit-switched and packet-switched traffic as described in Sections 4.3.4 and 4.3.5, respectively. The calculation is done for each

Table 8.9 Required circuit-switched system capacity per cell for RATG 1 in the dense urban teledensity in the year 2020 (kbit/s/cell).

Service category	RATG 1		
	Macro cell	Micro cell	Pico cell
SC 1	0	0	0
SC 2	0	**73 280.00**	**24 288.00**
SC 3	0	8496.00	1744.00
SC 4	5392.00	6752.00	368.00
SC 5	1792.00	3584.00	48.00
SC 6	0	0	0
SC 7	**37 904.00**	68 832.00	20 646.00
SC 8	15 824.00	44 670.24	2656.00
SC 9	0	7104.00	400.00
SC 10	0	3584.00	48.00

Table 8.10 Required circuit-switched unicast system capacity per cell for RATG 2 in the dense urban teledensity in the year 2020 (kbit/s/cell).

Service category	RATG 2			
	Macro cell	Micro cell	Pico cell	Hot spot
SC 1	0	0	0	0
SC 2	0	**321 920.00**	42 976.00	22 912.00
SC 3	0	211 072.00	3968.00	848.00
SC 4	60 384.00	119 616.00	2576.00	336.00
SC 5	21 312.00	27 808.00	208.00	48.00
SC 6	0	0	**354 560.00**	**332 672.00**
SC 7	**141 648.00**	312 992.00	39 472.00	19 456.00
SC 8	107 024.00	242 176.00	4592.00	1840.00
SC 9	0	138 720.00	2624.00	368.00
SC 10	0	27 808.00	208.00	48.00

service category. The resulting system capacity provides the QoS required for that particular service category. Thus, the maximum over all considered service categories defines the actual capacity requirement.

In Tables 8.9 and 8.10, the required system capacity for circuit-switched unicast traffic in the dense urban teledensity is shown for RATG 1 and RATG 2, respectively. Table 8.11 shows the required system capacity for the circuit-switched multicast traffic. Tables 8.12 and 8.13 show the required system capacity for packet-switched unicast traffic in the dense urban teledensity for RATG 1 and RATG 2, respectively. In each combination of RATG and radio

Table 8.11   Required circuit-switched multicast system capacity per cell in the dense urban teledensity in the year 2020 (kbit/s/cell).

Service category	RATG 1			RATG 2			
	Macro	Micro	Pico	Macro	Micro	Pico	Hot spot
SC 1	0	0	0	0	0	0	0
SC 2	**203 312.00**	0	0	**203 312.00**	0	0	0
SC 3	6624.00	0	0	6624.00	0	0	0
SC 4–10	0	0	0	0	0	0	0

environment, the service category that requires the maximum capacity is marked in bold font face. For the circuit-switched unicast traffic, the capacity demand is dominated by SC 7 in RATG 1 and RATG 2 macro cells, by SC 2 in RATG 1 micro cells, RATG 2 micro cells and RATG 1 pico cells, and by SC 6 in RATG 2 pico cells and hot spots. Circuit-switched multicast traffic is only offered for SC 2 and SC 3; the corresponding capacity demand is dominated by SC 2. For packet-switched unicast traffic, the dominating service categories are the same for the radio environments that exist in both RATGs. The macro cell capacity demand is dominated by SC 17, the micro cell by SC 12, the pico cell by SC 16, and the hot spot by SC 16 as well (RATG 1 does not have the hot spot radio environment deployed).

After determining the individual capacity demand for the uplink/downlink, unicast/multicast and circuit-switched/packet-switched traffic, these components are summed up to determine the total capacity requirement per combination of RATG and radio environment. Figures 8.7 and 8.8 show the total capacity demand by radio environment and the individual contributions to it by the packet-switched and circuit-switched service categories for uplink/downlink unicast and multicast traffic. Note that this summation is not a part of the methodology flow chart in Figure 4.2 but is shown for illustration only.

With a total of 472 Mbit/s the macro cell capacity demand is the smallest of the three RATG 1 radio environments. For the micro cells the capacity of 615 Mbit/s is required, and for the pico cells the capacity of 822 Mbit/s is needed.

The RATG 1 macro cell capacity requirement is mainly constituted of circuit-switched multicast downlink and packet-switched unicast uplink contributions. The overall uplink/downlink ratio is approximately 3:2. While circuit-switched unicast traffic shows a clear dominance of downlink capacity requirement over the uplink, the situation is the reverse for packet-switched traffic, where the uplink capacity requirement is almost three times the downlink capacity requirement.

RATG 1 only has multicast traffic in the macro cells. Consequently, the multicast contribution to capacity demand in micro and pico cells is zero. In both micro and pico cells the capacity demands are clearly dominated by packet-switched traffic. For the packet-switched traffic, the uplink/downlink ratio shows an emphasis of downlink traffic, while circuit-switched traffic capacity demand is rather symmetric.

For RATG 2 the required macro cell capacity is significantly smaller than for the other four radio environments. The highest capacity demand within RATG 2 occurs for the micro cells, whereas the pico cells and the hot spots have capacity requirements of about the same amount.

Table 8.12 Required packet-switched system capacity per cell for RATG 1 in the dense urban teledensity in the year 2020 (kbit/s/cell).

Service category	RATG 1		
	Macro cell	Micro cell	Pico cell
SC 11	0	0	0
SC 12	0	**300 758.40**	157 158.52
SC 13	6511.89	141 499.83	18 400.72
SC 14	2490.97	129 053.18	13 588.25
SC 15	5911.94	183 101.93	26 684.25
SC 16	0	0	**384 721.35**
SC 17	**53 682.07**	192 427.17	60 725.31
SC 18	0	132 670.24	14 769.42
SC 19	2073.28	124 745.62	11 861.13
SC 20	0	107 540.45	7673.50

Table 8.13 Required packet-switched system capacity per cell for RATG 2 in the dense urban teledensity in the year 2020 (kbit/s/cell).

Service category	RATG 2			
	Macro cell	Micro cell	Pico cell	Hot spot
SC 11	0	549 749.00	204 893.21	180 193.78
SC 12	0	**1 152 026.16**	194 673.63	151 256.71
SC 13	12 438.92	1 000 611.63	49 463.60	12 796.77
SC 14	9472.38	967 657.19	42 105.30	8467.23
SC 15	19 144.98	1 131 036.76	69 858.02	17 810.59
SC 16	0	0	**413 230.09**	**380 856.44**
SC 17	**57 216.07**	1 135 713.22	90 144.73	56 891.61
SC 18	0	1 003 010.63	43 630.43	9 875.93
SC 19	8423.95	980 215.62	39 105.52	7258.13
SC 20	0	928 354.31	30 264.18	4255.80

As observed in RATG 1 above, also for the RATG 2 macro cells the required packet-switched uplink capacity is three times the required packet-switched downlink capacity. In the other three radio environments, no multicast traffic is present and thus there is no corresponding capacity requirement shown for micro cells, pico cells and hot spots. The micro cell capacity requirement is dominated by packet-switched downlink traffic; the uplink/downlink asymmetry reaches a value of almost 1:6. Also for circuit-switched traffic there is a considerable asymmetry of approximately 1:2 (different from the corresponding situation in RATG 1). In pico cells and hot spots we again have a different overall picture; both packet-switched and circuit-switched capacity requirements are rather symmetric, and also the contributions of packet-switched and circuit-switched service categories to the total capacity requirement are rather equal in size.

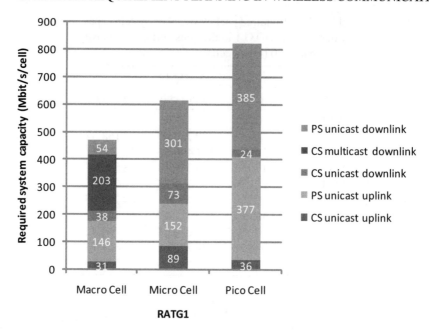

Figure 8.7 Required total system capacity for RATG 1 in dense urban teledensity in the year 2020 (Mbit/s/cell).

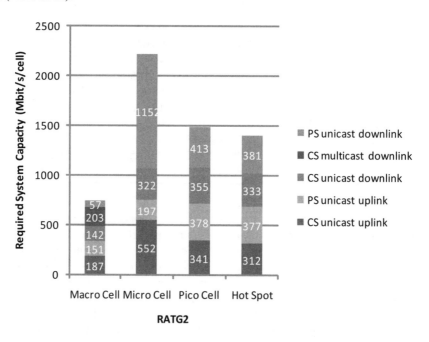

Figure 8.8 Required total system capacity for RATG 2 in dense urban teledensity in the year 2020 (Mbit/s/cell).

8.6 Required Spectrum

The initial spectrum requirement, i.e. the spectrum needed to provide the required system capacity is determined by dividing the capacity requirement by the spectral efficiency as described in Section 4.3.6. This is done in the sixth step of the methodology flowchart in Figure 4.2. The results from this operation are shown in Figure 8.9 for RATG 1 and

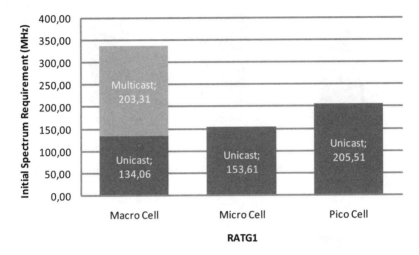

Figure 8.9 Initial spectrum requirement for RATG 1 in the dense urban teledensity in the year 2020 (MHz/cell).

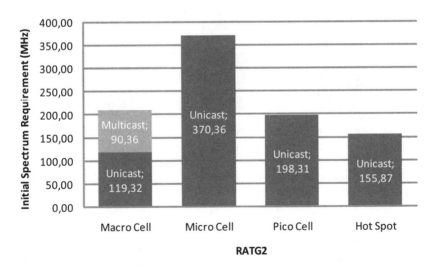

Figure 8.10 Initial spectrum requirement for RATG 2 in the dense urban teledensity in the year 2020 (MHz/cell).

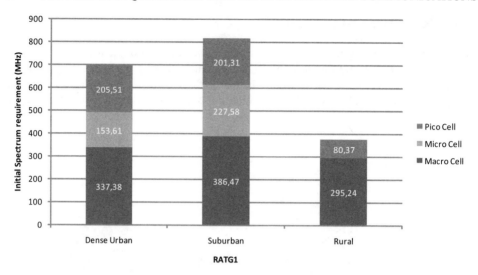

Figure 8.11   Variation over teledensities of initial spectrum requirement for RATG 1 in the year 2020 (MHz/cell).

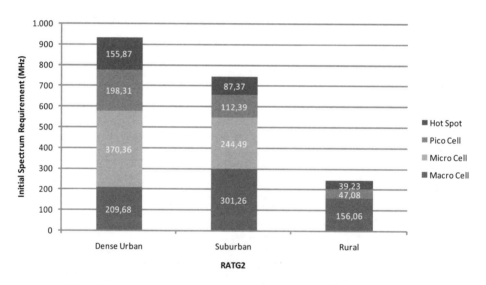

Figure 8.12   Variation over teledensities of initial spectrum requirement for RATG 2 in the year 2020 (MHz/cell).

Figure 8.10 for RATG 2. The initial spectrum requirements are shown as a total figure including both packet-switched and circuit-switched service categories in uplink and downlink directions. Multicast traffic is only present in the macro cells of both RATG 1 and RATG 2. In the RATG 1 macro cells the spectrum required by multicast traffic is higher than that by unicast traffic, while for RATG 2 the situation is the other way round.

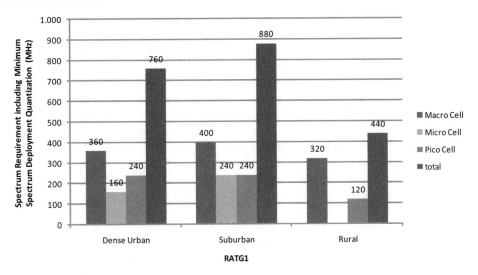

Figure 8.13 Spectrum requirement after adjustment with the minimum spectrum deployment for RATG 1 in the year 2020 (MHz/cell).

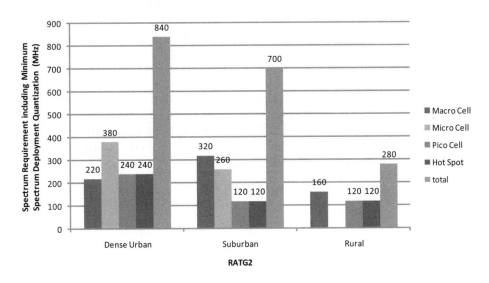

Figure 8.14 Spectrum requirement after adjustment with the minimum spectrum deployment for RATG 2 in the year 2020 (MHz/cell).

Although the RATG 1 capacity requirements for micro cells and pico cells are significantly higher than that for macro cells, the corresponding initial spectrum demand does not show the same relation due to the significantly higher spectral efficiency of micro and pico cells in RATG 1. For RATG 2 the highest initial spectrum demand occurs for the micro cell radio environment. Macro and pico cells require approximately the same spectrum, but in the macro cells almost half of the demand is due to multicast traffic. The hot spot radio environment requires less spectrum than the pico cells.

Figures 8.11 and 8.12 show the variation of spectrum requirements over the three different teledensities. For RATG 1 the highest spectrum demand occurs in the suburban teledensity, while for RATG 2 the maximum spectrum is required in the dense urban teledensity. Both RATGs show a higher spectrum demand in macro cells for suburban areas than for dense urban areas. While in RATG 1 the spectrum requirement in macro and micro cells grows from dense urban to suburban areas, in RATG 2 the demand only in the macro cell increases from dense urban to suburban areas.

The initial spectrum requirement has to be rounded up to the next integer multiple of the assumed minimum spectrum deployment in order to obtain a realistic result. This is done in the seventh step of the methodology flowchart in Figure 4.2. The result of this rounding process is shown in Figure 8.13 for RATG 1. Figure 8.14 shows the corresponding result for RATG 2. In both figures the required total amount is shown in addition to the requirements of the individual radio environments in each column group.

Note that RATG 1 micro cells are not assumed to be deployed in rural areas. For RATG 1 the spectrum demand including the minimum spectrum deployment quantization is dominated by the suburban teledensity, while for RATG 2 the dense urban teledensity dominates the spectrum demand. Within the respective dominating teledensities the macro cells give the largest contribution to the RATG 1 spectrum demand, while the micro cells represent the highest contribution to the RATG 2 spectrum demand.

Note that the total spectrum requirement per teledensity for RATG 2 is the sum over macro and micro cells plus the maximum of pico cells and hot spots, since the latter two radio environments are assumed to be spatially non-overlapping, allowing ideal spectrum reuse between pico cells and hot spots.

The finally resulting spectrum demands are 880 MHz for RATG 1 and 840 MHz for RATG 2 as the maximum of the different teledensities for each RATG, respectively, leading to a total spectrum requirement of 1720 MHz in the year 2020. We note that these values were actually reported as the spectrum requirements to the World Radiocommunication Conference 2007 (WRC-07). See Chapter 10.

9

Capacity Dimensioning to Meet Delay Percentile Requirements

Tim Irnich and Bernhard H. Walke

Current standards for IMT-2000 systems specify the Quality of Service (QoS) perceived by an application running on a mobile terminal by performance measures such as mean throughput, mean delay and percentage of packets for which a specified delay value must not be exceeded (3GPP 2006). The user is assumed to be 'satisfied' if the values specified for the respective measures are met on average. Averaging is done by taking all mobile terminals roaming in a cell into account. Apparently, terminals roaming close to a base station have a much better chance to be satisfied than terminals roaming close to the cell border. The methodology introduced in Chapter 4 takes two measures into account, namely, throughput and mean delay and presents a method to calculate the spectrum required to meet the required values for these two measures for IMT-Advanced systems. As mentioned above, a large variance in satisfaction is expected in relation to the current location of a mobile terminal. In particular, real-time applications can be expected to be served with dissatisfaction by a system dimensioned in its capacity only to meet the mean delay value. Therefore, the percentage of events with which a specified delay value is exceeded should also be taken into account and the system capacity should be calculated under this additional constraint.

In this chapter we extend the method to calculate the required system capacity for packet-switched service categories described in Section 4.3.5. We consider so-called *packet delay percentiles* and their impact on the system capacity. An *r*th *percentile* with respect to the probability distribution of a random variable X is defined as the value of X that is not exceeded with probability r. In standards for current mobile radio systems, mostly the 95th percentile is specified for real-time services, but any other value could be applied in the method described in the following.

Spectrum Requirement Planning in Wireless Communications   Edited by Hideaki Takagi and Bernhard H. Walke
© 2008 John Wiley & Sons, Ltd

This chapter is organized as follows. Section 9.1 describes a framework of the capacity dimensioning method based on the delay percentile. Section 9.2 introduces and motivates the multi-modal distribution of the service times in Internet Protocol (IP)-based mobile communication systems. In Section 9.3, the properties of the distribution function (DF) for the waiting time in an M/G/1 queue with multi-modal service time distribution are first discussed. Then the influence of nonpreemptive priority service discipline on the waiting time DF is evaluated. Finally, two approaches to approximate the waiting time DF are proposed. Section 9.4 shows how the delay DF and its percentiles can be obtained from approximations of the waiting time DF and the service time DF. In Section 9.5, the accuracy of the approximation for the delay DF is evaluated. Section 9.6 presents results showing the impact of considering delay percentiles as QoS requirements on the required system capacity. Conclusions are drawn in Section 9.7.

We note that the capacity dimensioning based on the delay percentile is not included in the ITU-R methodology for IMT-Advanced.

9.1   Delay Percentile Evaluation

Let us denote by $\mathcal{D}_n$ a random variable that represents the *delay* of a packet of class n. In our terminology, the delay consists of the *waiting time*, later denoted by $\mathcal{W}_n$, and the *service time* denoted by $\mathcal{T}_n$. We focus on the DF for the delay $\mathcal{D}_n$ which is the probability that $\mathcal{D}_n$ does not exceed t, $P\{\mathcal{D}_n \leq t\}$, as a function of t. The rth percentile of the DF for the delay $\mathcal{D}_n$ is given as the value of t such that

$$P\{\mathcal{D}_n \leq t\} = r.$$

The required rth percentile of the delay is denoted by $\pi_n(r)$. This implies that the inequality

$$P\{\mathcal{D}_n \leq \pi_n(r)\} \geq r \tag{9.1}$$

must be satisfied (Figure 9.1). The delay DF is a decreasing function of the system capacity C (bits/s), which means that the delay percentile decreases if C is increased. Suppose that the initial value of C is given somehow, for example, by the methodology based on the mean delay requirement as described in Section 4.3.5. If the condition in Equation (9.1) is not satisfied with this value of the system capacity, we increase the capacity by a small amount and then check if it is satisfied with the new value. We repeat this procedure until the condition in Equation (9.1) is satisfied. The resulting system capacity is the one that meets the required rth percentile of the delay. This iterative procedure is described as a flow chart in Figure 9.2.

A central element of this approach is the determination of the packet delay percentile under a given system load and system capacity. In general the delay DF in a queueing system can be determined either analytically or by simulation. However, for repeated calculations of a delay percentile in our iterative dimensioning procedure, the simulation is not appropriate because it takes too much computational time. For M/G/1-type queues, the exact distribution of the delay is given in the *Laplace–Stieltjes transform* (LST) of the DF

$$D_n^*(s) := \int_0^\infty e^{-st}\, dP\{\mathcal{D}_n \leq t\}.$$

Hence, analytical determination of the delay DF requires inversion of the LST. If only a certain value of the DF is desired, a numerical method for inverting the LST is possible.

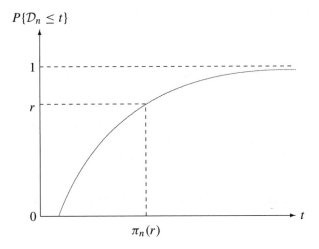

Figure 9.1 The rth percentile of the distribution function $P\{\mathcal{D}_n \leq t\}$.

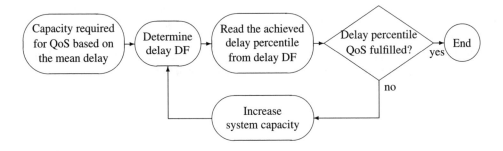

Figure 9.2 Flow chart of the procedure for calculating the delay percentile-bound system capacity.

However, available algorithms for numerical transform inversion are not suitable for the purpose of determining percentiles, which is reading the DF $P\{\mathcal{D}_n \leq t\}$ at a given probability value instead of reading the DF at a certain value of t. We therefore focus on analytical determination of the delay DF. Unfortunately, closed-form transform inversions are not available in the literature for the delay DF or the waiting time DF in a single-class M/G/1 queue with first-come first-served (FCFS) service discipline, let alone in an M/G/1 nonpreemptive priority queue used in the mean delay-based methodology introduced in Section 4.3.5. Thus we apply an approximation technique to obtain the delay DF. It is assumed that the mean and the second moment of the delay are known. Then the approximation is based on selecting the parameters in the approximate form for the delay DF so that the mean and the second moments of this DF match the known mean and the second moment, respectively, of the delay.

In order to utilize known properties of service time distributions in packet-based communication systems, we determine the delay DF indirectly by first determining the waiting time DF and then obtaining the delay DF by the convolution with the probability density function (pdf) for the service time. Recall that the DF of the sum of two independent random variables can be obtained through the convolution of the DF of one random variable with the pdf of the other random variable. In the present case, the delay is the sum of the waiting time and the service time that are mutually independent.

9.2 Service Time Distribution in IP-Based Communication Systems

Data packets in communication systems contain an integer number of octets, which means that the packet length (counted in bytes) can assume only discrete values. Since we are focusing on IP packets, the packet size is bounded by the Maximum Transfer Unit (MTU) of IP (Stevens 1994). IP packet size is a discrete random variable with a limited range. The minimum IP packet size without header compression is 20 bytes, and the maximum size is determined by the MTU. Furthermore, measurements show that the service time distributions in real packet-based communication systems are dominated by a few (usually three or four) accumulation points or *modes* (Thompson *et al.* 1997). Therefore, the computational and parametrization effort can be limited to considering *multi-modal distributions* for the packet service time.

We assume that there are N classes of packets, and that the length of a packet of class n equals $x_{m,n}$ (bits) with probability $p_{m,n}$, $m = 1, \ldots, M_n$, such that

$$x_{1,n} < x_{2,n} < \cdots < x_{M_n,n} \quad \text{and} \quad \sum_{m=1}^{M_n} p_{m,n} = 1$$

for $n = 1, 2, \ldots, N$. If packets are transmitted over a channel with capacity C (bit/s), the transmission time (s) of a packet of length $x_{m,n}$ is $t_{m,n} = x_{m,n}/C$. This leads to the following pdf for the service time T_n of a packet of class n:

$$P\{T_n = t\} = \sum_{m=1}^{M_n} p_{m,n} \delta(t - t_{m,n}), \tag{9.2}$$

where $\delta(t)$ denotes the Dirac impulse at $t = 0$.

We denote by 'M/D*/1' an M/G/1 queue with this multi-modal distribution of the service times. Hence a special challenge arises for finding a suitable approximation for the waiting time DF in M/D*/1 queues. The desired waiting time DF approximation will be used to determine the high order percentiles of the waiting time distribution, e.g. the 95th percentile. Therefore, we are particularly interested in finding approximations that are accurate in the tail of the waiting time DF.

In this case the first three moments of the packet size distribution for class n are

$$S_n = \sum_{m=1}^{M_n} p_{m,n} x_{m,n}; \quad s_n^{(2)} = \sum_{m=1}^{M_n} p_{m,n} x_{m,n}^2; \quad s_n^{(3)} = \sum_{m=1}^{M_n} p_{m,n} x_{m,n}^3. \tag{9.3}$$

The mean W_n and the second moment $W_n^{(2)}$ of the waiting time in an M/G/1 nonpreemptive priority queue are given in terms of the first three moments of the service time distribution of each class. Let λ_n be the arrival rate (packets/s) of packets of class n. Then we have

$$W_n = \frac{\sum_{i=1}^{N} \lambda_i s_i^{(2)}}{2(C - \sum_{i=1}^{n-1} \lambda_i s_i)(C - \sum_{i=1}^{n} \lambda_i s_i)} \tag{9.4}$$

and

$$W_n^{(2)} = \frac{\sum_{i=1}^{N} \lambda_i s_i^{(3)}}{3(C - \sum_{i=1}^{n-1} \lambda_i s_i)^2(C - \sum_{i=1}^{n} \lambda_i s_i)}$$

$$+ \frac{(\sum_{i=1}^{n} \lambda_i s_i^{(2)})(\sum_{i=1}^{N} \lambda_i s_i^{(2)})}{2(C - \sum_{i=1}^{n-1} \lambda_i s_i)^2(C - \sum_{i=1}^{n} \lambda_i s_i)^2}$$

$$+ \frac{(\sum_{i=1}^{n-1} \lambda_i s_i^{(2)})(\sum_{i=1}^{N} \lambda_i s_i^{(2)})}{2(C - \sum_{i=1}^{n-1} \lambda_i s_i)^3(C - \sum_{i=1}^{n} \lambda_i s_i)}. \tag{9.5}$$

We note that Equation (9.4) is derived from Equation (A.89) in Appendix A.4 as was done for Equation (4.28). The second moment of the waiting time in an M/G/1 nonpreemptive priority queue was first obtained by Kesten and Runnenberg (1957). Its derivation can be found in Takagi (1991, p. 293). The mean W_n and the second moment $W_n^{(2)}$ will be used in the approximations for the waiting time DF in the next section.

9.3 Waiting Time Distribution in M/G/1 Queues

The LST of the waiting time DF in an M/G/1-FCFS queue is given by the well-known *Pollaczek–Khintchin transform*:

$$W^*(s) = \frac{(1-\rho)s}{s - \lambda + \lambda B^*(s)}, \tag{9.6}$$

where λ is the packet arrival rate (packets/s), $\rho = \lambda s$ is the system load (dimensionless) with s being the mean service time (s/packet), and $B^*(s)$ is the LST of the DF for the service time. This formula is shown in standard textbooks on queueing theory such as Cooper (1991, p. 217), Fujiki and Gambe (1980, p. 284), Gross and Harris (1998, p. 226), Kleinrock (1975, p. 200), and Syski (1986, p. 198).

Obviously a closed-form transform inversion of the waiting time LST requires the service time LST to be known, and the resulting expression needs to be analytically tractable.

9.3.1 Waiting time under multi-modal service time distribution

Since we aim at multi-modal service time distributions it is natural to check if this special case allows the transform inversion to be performed analytically.

The simplest multi-modal service time distribution is the case in which the service time is a constant D. The DF for the waiting time in an M/D/1-FCFS queue is given by

$$W(t) = (1 - \rho) \sum_{i=0}^{\lfloor t/D \rfloor} \frac{(i\rho - \lambda t)^i}{i!} e^{-(i\rho - \lambda t)}, \tag{9.7}$$

where λ is the packet arrival rate, $\rho = \lambda D$ is the system load and $\lfloor t/D \rfloor$ is the floor function giving the largest integer not exceeding t/D. This expression was obtained by A. K. Erlang in 1909; see Brockmeyer *et al.* (1948, p. 133) and Syski (1986, p. 220).

An extension to a discrete service time distribution with multiple possible values is derived by Shortle and Brill (2005). In their framework the possible service times are integer multiples of a constant D. Shortle and Brill denote this queue by M/{iD}/1. In their model each service time equals iD with probability p_i, where $\sum_{i=1}^{\infty} p_i = 1$, and $\lambda_i = p_i \lambda$ is the arrival rate of customers that have service time iD, $i = 1, 2, \ldots$. Then the waiting time DF of an M/{iD}/1-FCFS queue is given by

$$W(t) = (1 - \rho) e^{\lambda t} + (1 - \rho) \sum_{i=1}^{\lfloor t/D \rfloor} e^{-\lambda(iD-t)} \sum_{\mathcal{L} \in \mathcal{P}(i)} \frac{(iD - t)^{|\mathcal{L}|}}{H(\mathcal{L})} \prod_{j \in \mathcal{L}} \lambda_j. \tag{9.8}$$

The corresponding pdf is given by

$$w(t) = (1 - \rho)\lambda e^{\lambda t} + (1 - \rho) \sum_{i=1}^{\lfloor t/D \rfloor} e^{-\lambda(iD-t)}$$

$$\times \sum_{\mathcal{L} \in \mathcal{P}(i)} \{\lambda(iD - t) - |\mathcal{L}|\} \frac{(iD - t)^{|\mathcal{L}|-1}}{H(\mathcal{L})} \prod_{j \in \mathcal{L}} \lambda_j. \tag{9.9}$$

Here, $\mathcal{P}(i)$ is the set of all possible partitions of positive integer i. A partition of i is the set of the combination of positive integers that sum to i. For example

$$\mathcal{P}(4) = \{\{4\}, \{3, 1\}, \{2, 2\}, \{2, 1, 1\}, \{1, 1, 1, 1\}\}.$$

Furthermore, $H(\mathcal{L})$ is a function defined as follows. Let $\mathcal{L}$ be a partition in $\mathcal{P}(i)$. Let $r_1 > r_2 > \cdots > r_d$ be the d *distinct* integers in the partition with $n_1, n_2, \ldots, n_d$ being the respective multiplicities so that $\sum_{k=1}^{d} n_k r_k = i$. For example, if $\mathcal{L} = \{4, 2, 2, 2\} \in \mathcal{P}(10)$, then $r_1 = 4, r_2 = 2, n_1 = 1, n_2 = 3$ and $d = 2$. Then, $H(\mathcal{L})$ is defined as

$$H(\mathcal{L}) := n_1! n_2! \cdots n_d!.$$

Using the same notation, we write

$$|\mathcal{L}| := n_1 + n_2 + \cdots + n_d$$

and

$$\prod_{j \in \mathcal{L}} \lambda_j = \lambda_{r_1}^{n_1} \lambda_{r_2}^{n_2} \cdots \lambda_{r_d}^{n_d}.$$

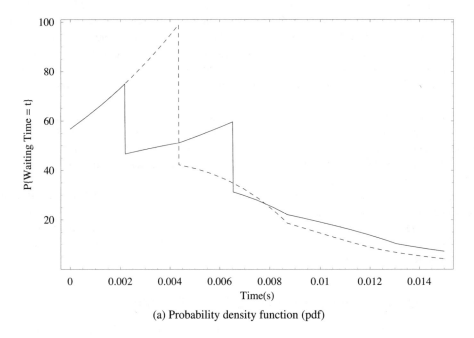

(a) Probability density function (pdf)

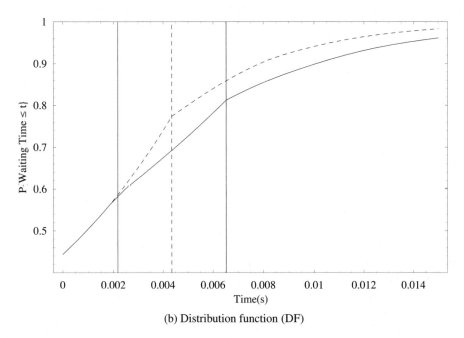

(b) Distribution function (DF)

Figure 9.3 Waiting time distribution in M/{iD}/1 queues with one- and two-point service time distributions (dotted and solid lines, respectively).

In Figure 9.3, the pdf and DF for the waiting time are shown for an $M/\{i\,D\}/1$-FCFS queue with the offered traffic of 500 kbit/s and the mean packet size of 500 bytes. Two different packet size distributions are assumed. The first example is a deterministic, i.e. one-point distribution where all packets have a length of 500 bytes. The second example is a two-point distribution where 50% of all packets contain 250 bytes and the other 50% have the length of 750 bytes. Since the offered traffic is the average number of bits that arrive per second, the packet arrival rate in both examples is $\lambda = 125$ packets/s.

The examples in Figure 9.3 show that the waiting time DF has a probability mass at $t = 0$. It corresponds to the probability of immediate access to the server, i.e. $W(0) = 1 - \rho$ (Equation (9.8)). The pdf has discontinuities located at $t = i\,D$ for all i such that $p_i \neq 0$, i.e. at the locations of different possible values of the service time. The height of the drop at the ith discontinuity is $\lambda(1 - \rho)p_i$.

Based on these observations we can expect that the waiting time DF can be difficult to approximate with a continuous function if the region of interest is close to one of the discontinuities. In particular, if there are only a few possible service time values, such an approximation may be inaccurate, since in such cases the drops in the pdf are quite high, leading to relatively 'sharp' corners in the DF.

9.3.2   Influence of nonpreemptive priority discipline

Before discussing specific approximation techniques for the waiting time distribution, we evaluate the influence of the nonpreemptive priority discipline on the waiting time distribution properties. There are no closed-form solutions for the waiting time distribution in an $M/D^*/1$ nonpreemptive priority queue available in the literature, so we have to use simulation results for these evaluations.

In Figure 9.4 the waiting time DF of an $M/D^*/1$-FCFS queue is compared with an equally loaded $M/D^*/1$ nonpreemptive priority queue with three classes. The total arrival rate in both queues is 375 packets/s, with equal traffic of 125 packets/s in each class. In Figure 9.4(a), it is assumed that all packets have the same length of 500 bytes and that the server speed is 1.914 Mbit/s. In Figure 9.4(b), a two-point distribution is assumed such that 50% of packets contain 250 bytes and the other 50% of packets 750 bytes, and the server speed is 1.979 Mbit/s. The locations of the respective service time values are marked with vertical lines for easier comparison of the corresponding DF values at these locations.

We observe in Figure 9.4 that the introduction of nonpreemptive priority discipline causes the discontinuities in the DF's first derivative (i.e. the 'corners') to move to higher probability values for the high and medium priority classes than in the FCFS queue. This implies that determining the waiting time or delay percentiles at high probability values from an approximation of such DF may be more problematic in the presence of priority discipline than in the $M/D^*/1$-FCFS queue. Furthermore, the drops of the probability density function become higher (which can be seen from the 'sharper' corners of the DF) adding to the difficulties of approximating these particular regions of the DF. In contrast, the DF for the low-priority class in the priority queue resides below that for the FCFS queue. Therefore, the introduction of priority discipline makes the approximation of high order percentiles for the low-priority class easier than for the $M/D^*/1$-FCFS queue.

In Figures 9.5 and 9.6, the SCV for the waiting times of all classes and the pdf for the waiting time of the high priority class are shown for $M/D^*/1$ nonpreemptive priority queues

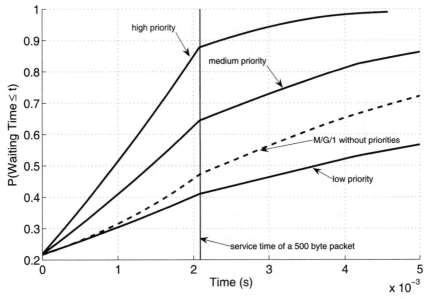

(a) One-point service time distribution at 500 bytes with server capacity of 1.914 Mbit/s

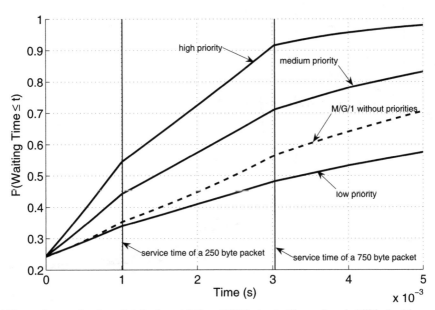

(b) Two-point service time distribution at 250 and 750 bytes (with equal probabilities) with server capacity of 1.979 Mbit/s

Figure 9.4 Comparison of M/D*/1 waiting time DFs for strict FCFS and nonpreemptive priority service (FCFS: analysis, nonpreemptive priority: simulation).

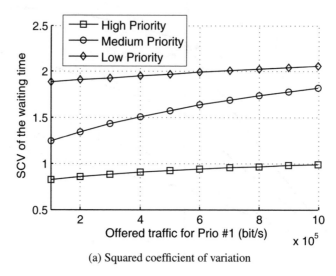

(a) Squared coefficient of variation

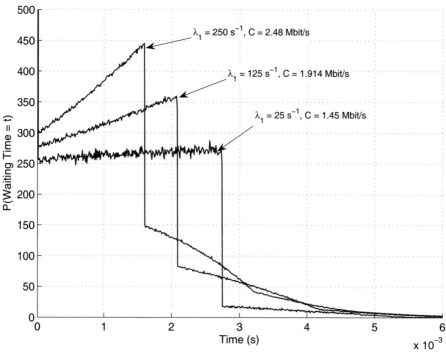

(b) Probability density function for the waiting time of the high priority class

Figure 9.5 SCV and pdf of the waiting times in a system of three priority classes with one-point packet size distribution at 500 bytes, generated by simulation.

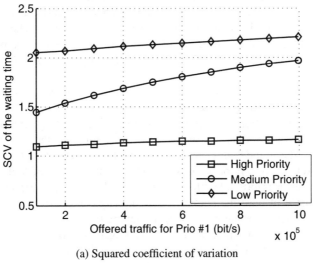

(a) Squared coefficient of variation

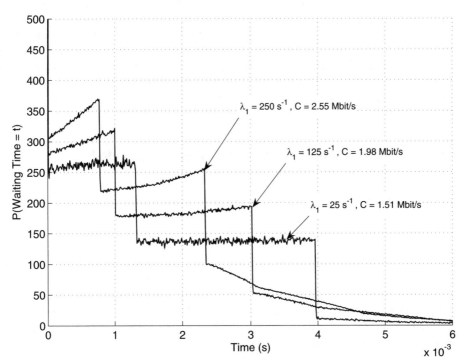

(b) Probability density function for the waiting time of the high priority class

Figure 9.6 SCV and pdf of the waiting times in a system of three priority classes with two-point packet size distribution at 250 bytes and 750 bytes with equal probabilities, generated by simulation.

with three priority classes, resulting from a simulation model. For the medium and low priority classes, the offered traffic is assumed to be fixed at 500 kbit/s, and the offered traffic for the high priority class is varied. In Figure 9.5, a deterministic packet size distribution with packets of 500 bytes is assumed for all classes. In Figure 9.6, a two-point packet size distribution with 50% of packets of 250 bytes and 50% of packets of 750 bytes is assumed for all classes.

Obviously, waiting time distributions that have an SCV smaller than one may occur in M/D*/1 nonpreemptive priority queues. The highest priority class seems to have the greatest likelihood of such a situation. Factors that may reduce the waiting time SCV in the high priority class are the low variance of service time distributions in the low priority class and the low offered traffic in the high priority class. The latter leads to a significant probability for arriving packets of high priority class to find an empty queue of high priority packets with a packet of lower priority class being served. In this case, the arriving packet of high priority class must wait until the packet of low priority class currently served is finished, which significantly decreases the variance of its waiting time.

In Figure 9.5(b), we can observe that the waiting time pdf of the high priority class resembles a uniform distribution; the jumps are located at the service time of 500 bytes. In Figure 9.6(b), we can see that each mode of the service time distributions of the lower priority classes contributes an additional part to the waiting time pdf, which in case of low offered traffic in the high priority class comes close to a uniform distribution. Each uniform-like part of the distribution has its drop at the corresponding service time, in this example at the service times of a 250 and 750 byte packet, respectively. The lower border of the SCV high priority class is thus strongly influenced by the number of different possible packet size values over all present priority classes. The more diverse the arriving traffic mix is, the lower is the probability of SCV values below one.

9.3.3 Waiting time approximation based on degenerated hyperexponential distribution

The DF for the *hyperexponential distribution* of the second order is given by

$$F(t) = 1 - p_1 e^{-\gamma_1 t} - p_2 e^{-\gamma_2 t}, \quad t \geq 0 \tag{9.10}$$

where $p_1 + p_2 = 1$. We consider a special case in which $\gamma_1 = \infty$ in order to create a probability mass at $t = 0$. This special case is called *degenerated hyperexponential distribution* of the second order, which we denote by H_2. It is equivalent to the sum of a p_1-weighted Dirac impulse at $t = 0$ and a p_2-weighted exponential distribution with rate γ_2.

We may use the following degenerated hyperexponential distribution of the second order to approximate the waiting time DF of service class n:

$$P\{\mathcal{W}_n \leq t\} = 1 - \rho_n e^{-\gamma_n t}, \quad t \geq 0. \tag{9.11}$$

If the parameters ρ_n and γ_n are determined according to

$$\gamma_n = \frac{2W_n}{W_n^{(2)}} \quad \text{and} \quad \rho_n = \frac{2W_n^2}{W_n^{(2)}}, \tag{9.12}$$

where the exact mean W_n and the exact second moment $W_n^{(2)}$ are given in Equations (9.4) and (9.5), respectively. Then the mean of Equation (9.11) is equal to W_n and the second moment is equal to $W_n^{(2)}$.

$$E[W_n] = \int_0^\infty P\{W_n > t\}\,dt = \rho_n \int_0^\infty e^{-\gamma_n t}\,dt = \frac{\rho_n}{\gamma_n} = W_n \qquad (9.13)$$

$$E[W_n^2] = 2 \int_0^\infty t\,P\{W_n > t\}\,dt = 2\rho_n \int_0^\infty t\,e^{-\gamma_n t}\,dt = \frac{2\rho_n}{\gamma_n^2} = W_n^{(2)}. \qquad (9.14)$$

Hence Equation (9.11) approximates the waiting time DF so that its first two moments match the mean and second moment of the waiting time DF.

Determination of the parameters γ_n and ρ_n only leads to valid results if

$$\rho_n \le 1 \quad \Leftrightarrow \quad 2\frac{W_n^2}{W_n^{(2)}} \le 1 \quad \Leftrightarrow \quad \frac{W_n^{(2)}}{W_n^2} \ge 2. \qquad (9.15)$$

The *squared coefficient of variation* (SCV) for the waiting time W_n is defined by

$$c_{W_n}^2 := \frac{\mathrm{Var}[W_n]}{W_n^2} = \frac{W_n^{(2)}}{W_n^2} - 1. \qquad (9.16)$$

Then Equation (9.15) can be written as

$$c_{W_n}^2 \ge 1. \qquad (9.17)$$

This condition states that the waiting time distribution to be approximated by Equation (9.11) must have a variance larger than that of an exponential distribution.

9.3.4   Waiting time approximation based on gamma distribution

As an alternative approximation, the gamma distribution can be parameterized to fit any kind of distribution function, although it is more difficult to handle.

Since no closed-form inverse of the gamma function is available, a numerical algorithm, e.g. Newton's method, must be applied to determine percentiles from the gamma distribution. This can significantly increase the computational effort. In particular, since we intend to use this approximation in an iterative procedure of percentile determination, comparison with requirements and increase of capacity, a significant optimization effort is required in order to achieve an acceptable execution time of the whole process of spectrum requirement calculation.

In this case, the function to be used for approximation of the waiting time DF is given by the *gamma distribution*

$$P\{W_n \le t\} = \frac{\gamma(k_n, t/\Theta_n)}{\Gamma(k_n)}, \quad t \ge 0, \qquad (9.18)$$

where

$$\Gamma(k) := \int_0^\infty x^{k-1}\,e^{-x}\,dx, \quad k > 0 \qquad (9.19)$$

is the gamma function. Note that $\Gamma(k) = (k-1)!$ for positive integer k. Also

$$\gamma(k, z) := \int_0^z x^{k-1} e^{-x} \, dx \tag{9.20}$$

is the lower incomplete gamma function (Abramowitz and Stegun 1972, p. 255 and p. 260). Note that $\gamma(k, \infty) = \Gamma(k)$.

If the parameters k_n and Θ_n are determined according to

$$k_n = \frac{W_n^2}{W_n^{(2)} - W_n^2} \quad \text{and} \quad \Theta_n = \frac{W_n^{(2)} - W_n^2}{W_n}, \tag{9.21}$$

then the mean of Equation (9.18) is equal to W_n and the second moment is equal to $W_n^{(2)}$. To see this, we first note that

$$P\{\mathcal{W}_n > t\} = 1 - P\{\mathcal{W}_n \leq t\} = 1 - \frac{\gamma(k_n, t/\Theta_n)}{\Gamma(k_n)}$$

$$= \frac{1}{\Gamma(k_n)} \int_{t/\Theta_n}^{\infty} x^{k_n-1} e^{-x} \, dx. \tag{9.22}$$

Then we have

$$E[\mathcal{W}_n] = \int_0^{\infty} P\{\mathcal{W}_n > t\} \, dt = \frac{1}{\Gamma(k_n)} \int_0^{\infty} dt \int_{t/\Theta_n}^{\infty} x^{k_n-1} e^{-x} \, dx$$

$$= \frac{1}{\Gamma(k_n)} \int_0^{\infty} x^{k_n-1} e^{-x} \, dx \int_0^{\Theta_n x} dt = \frac{\Theta_n}{\Gamma(k_n)} \int_0^{\infty} x^{k_n} e^{-x} \, dx$$

$$= k_n \Theta_n = W_n, \tag{9.23}$$

$$E[\mathcal{W}_n^2] = 2 \int_0^{\infty} t P\{\mathcal{W}_n > t\} \, dt = \frac{2}{\Gamma(k_n)} \int_0^{\infty} t \, dt \int_{t/\Theta_n}^{\infty} x^{k_n-1} e^{-x} \, dx$$

$$= \frac{2}{\Gamma(k_n)} \int_0^{\infty} x^{k_n-1} e^{-x} \, dx \int_0^{\Theta_n x} t \, dt$$

$$= \frac{\Theta_n^2}{\Gamma(k_n)} \int_0^{\infty} x^{k_n+1} e^{-x} \, dx = k_n(k_n + 1)\Theta_n^2 = W_n^{(2)}. \tag{9.24}$$

Hence Equation (9.18) approximates the waiting time DF so that its first two moments match the mean and second moment of the waiting time DF.

9.4   Delay DF Approximation

The packet delay over the air interface of packet-based wireless systems consists of the waiting time in a queue and the transmission time over the air interface. Denoting by $\mathcal{D}_n$ the packet delay random variable of priority class n, we have

$$\mathcal{D}_n = \mathcal{W}_n + \mathcal{T}_n. \tag{9.25}$$

Since $\mathcal{W}_n$ and $\mathcal{T}_n$ are mutually independent, the delay DF can be calculated as the convolution of the waiting time DF and the service time pdf. Using the two different waiting time approximations introduced in Sections 9.3.3 and 9.3.4, two different approximations for the delay DF can be obtained. In both cases the result of convolution can be determined in a straightforward manner, since the convolution with a Dirac impulse at $t = t_{m,n}$ is simply a shift of the argument by $t_{m,n}$.

Delay DF using H_2 waiting time approximation

Given the waiting time DF approximation in Equation (9.11) and service time pdf in Equation (9.2), the resulting delay DF for class n is calculated to be

$$P\{\mathcal{D}_n \leq t\} = \begin{cases} 0, & t < t_{1,n} \\ \displaystyle\sum_{m=1}^{i} p_{m,n}[1 - \rho_n\, e^{-\gamma_n(t - t_{m,n})}], & t_{i,n} \leq t < t_{i+1,n} \\ & \text{and } 1 \leq i \leq M_n - 1 \\ 1 - \rho_n \displaystyle\sum_{m=1}^{M_n} p_{m,n}\, e^{-\gamma_n(t - t_{m,n})}, & t \geq t_{M_n,n}, \end{cases} \tag{9.26}$$

where $t_{n,m}$ and $p_{n,m}$ $(m = 1, \dots, M_n)$ are the different possible service time values and their respective occurrence probabilities defined in Section 9.2.

Delay DF using gamma waiting time approximation

Given the waiting time DF approximation in Equation (9.18) and service time pdf in Equation (9.2), the resulting delay DF for class n is calculated to be

$$P\{\mathcal{D}_n \leq t\}$$
$$= \begin{cases} 0, & t < t_{1,n} \\ \displaystyle\sum_{m=1}^{i} p_{m,n} \frac{\gamma[k_n, (t - t_{m,n})/\Theta_n]}{\Gamma(k_n)}, & t_{i,n} \leq t < t_{i+1,n} \\ & \text{and } 1 \leq i \leq M_n - 1 \\ \displaystyle\sum_{m=1}^{M_n} p_{k,n} \frac{\gamma[k_n, (t - t_{m,n})/\Theta_n]}{\Gamma(k_n)}, & t \geq t_{M_n,n}. \end{cases} \tag{9.27}$$

9.5 Accuracy of Gamma and H_2 Approximations

In this section the accuracy of the waiting time approximation by means of H_2 and gamma functions is studied and compared to simulation results.

9.5.1 Approximation for high priority class

Since approximations accuracy is expected to be most error-prone for the high priority class (Section 9.3.2), we first focus on this priority class.

Waiting time DF

In Figure 9.7, the high priority waiting time DF with two corresponding approximations is shown for a system with three nonpreemptive priority classes. A bi-modal packet size distribution with packet lengths of 250 bytes and 750 bytes with equal occurrence probabilities is assumed. The offered traffic of the medium and low priority class is 500 kbit/s; the offered traffic T_1 for the high priority class is set to 10 kbit/s, 500 kbit/s and 1 Mbit/s, respectively. Figure 9.7(a) and (b) compares the H_2 and gamma approximations, respectively, to the simulation results. In each operating point the server capacity is dimensioned to meet the mean delay criterion of 20 ms for all three priority classes.

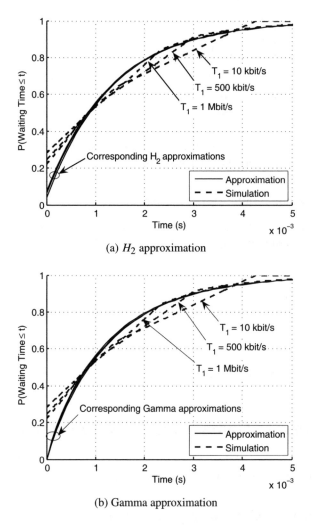

(a) H_2 approximation

(b) Gamma approximation

Figure 9.7 Comparison of H_2 and gamma waiting time DF approximations with simulation results for high priority class.

Consistently with the observations made in Section 9.3.3, for small offered traffic in the high priority class the discontinuity caused to the high priority waiting time DF by the 750 byte packets resides in the range of the 90 to 99th percentiles. With more offered traffic in the high priority class, the discontinuity moves down to lower probabilities, making both approximations more accurate. The more traffic is offered in the high priority class, the lower is the probability of an arriving high priority packet to find an empty high priority queue and the server occupied by a lower priority packet. Since the waiting time of such packets is limited by the maximum length of a lower priority packet, the waiting time DF of these packets degenerates towards a uniform distribution; see Section 9.3.3 for further explanation. We thus can conclude that the accuracy of both waiting time approximations primarily depends on the packet arrival rate of the respective priority class. The relation of high priority and lower priority packet arrival rates is a secondary influence factor. When the lower priority packet arrival rate is significantly larger than the high priority arrival rate, the dependency on the high priority packet arrival rate becomes stronger.

Delay DF

In Figure 9.8, the two delay DF approximations are shown for the same example system as used above with high priority offered traffic T_1 as the varying parameter.

Obviously the convolution of Equation (9.11)/(9.18) and Equation (9.2) removes the discontinuities of the resulting DF from the relevant percentiles in the range around the 95th percentile, which appears to be beneficial for the accuracy of the approximation. In addition, an increase of the high priority offered traffic load further improves the accuracy of the approximation. Both waiting time approximations, H_2 and gamma, appear to be comparable in precision to calculate percentiles of the delay DF.

Relative error

To perform a quantitative evaluation of the accuracy in the approximation, we use the *relative error* of the percentiles, which is defined by

$$\text{Relative error} := \frac{\text{Approximation's percentile} - \text{Simulation's percentile}}{\text{Simulation's percentile}}. \quad (9.28)$$

A positive relative error means that the approximation overestimates the actual (simulation) percentile, and a negative relative error means underestimation.

Figure 9.9 shows the relative error of the 80 to 99th percentiles in the delay DF of the high priority class for the same scenario as in Figure 9.8. A maximum error of below 15% is observed at the 99th percentile in the case of 10 kbit/s offered traffic in the high priority class; in all other cases the error is significantly less. This shows that both approximations are very accurate in this scenario. As expected from the observations made from Figure 9.8, additional high priority offered traffic significantly further reduces the error. For situations with not too low offered traffic in the high priority class, both approximation methods tend to overestimate the delay percentile in the most interesting range around the 95th percentile.

The relative error of the 95th percentile in the delay DF of the high priority class is plotted over a wide range of offered traffic values in Figure 9.10. In this figure, the H_2 approximation is found to be slightly more accurate than the gamma approximation. The peak observed in each curve comes from the largest packet's discontinuity moving across the 95th percentile

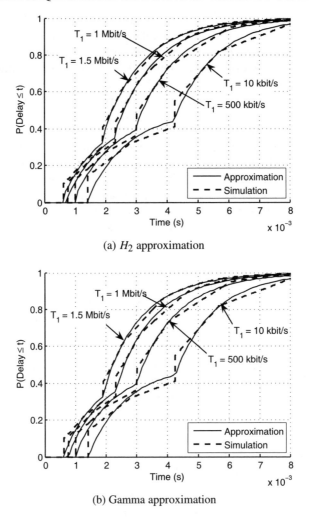

(a) H_2 approximation

(b) Gamma approximation

Figure 9.8 Comparison of H_2 and gamma delay DF approximations with simulation results for high priority class.

level when the offered traffic increases; see Figure 9.8. For further increased offered traffic we observe convergence towards a very low error such as values below 1%.

9.5.2 Approximation for medium and low priority classes

We now turn to assess the accuracy of the approximation for the medium and low priority classes.

In Figure 9.11, the delay DF approximations for the medium and low priority classes are compared with simulation. As in the above examples, the offered traffic for medium and low priority class is 500 kbit/s, and the high priority offered traffic T_1 is 10 kbit/s, 500 kbit/s and

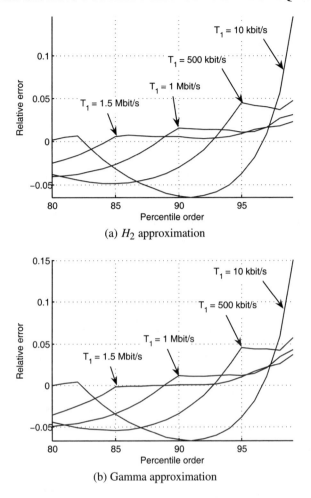

Figure 9.9 Relative error of actual and approximate delay percentiles for H_2 and gamma approximations for high priority class.

1 Mbit/s, respectively. In each operating point the server capacity is dimensioned to meet the mean delay criterion of 20 ms for all three classes.

While the discontinuity resulting from the largest packet size for high priority traffic is close to the relevant 95th percentiles, it resides at significantly lower probabilities for the medium and low priority delay DF. Thus we can conclude that the inaccuracy is mainly an issue for high priority classes; the lower the priority is, the less relevant this issue becomes. This means that the accuracy of approximation for lower priority classes is mainly determined by how well the selected functions match the tail probabilities of the waiting time DF. As already observed in Section 9.3.2 for the high priority delay DF, also in the medium and low priority delay DF the discontinuity associated with the maximum possible packet size value for lower priority classes moves towards lower probability, making approximation with continuous functions even more appropriate.

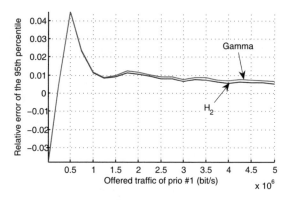

Figure 9.10 Relative error of the 95th percentile for H_2 and gamma approximations for high priority class with varying the offered traffic in that class.

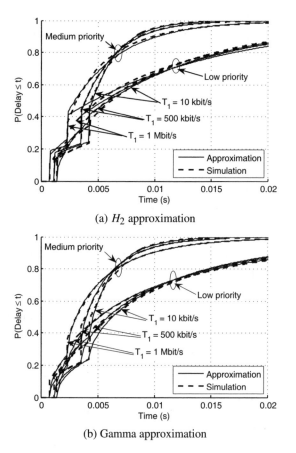

Figure 9.11 Comparison of H_2 and gamma delay DF approximations with simulation results for medium and low priority classes.

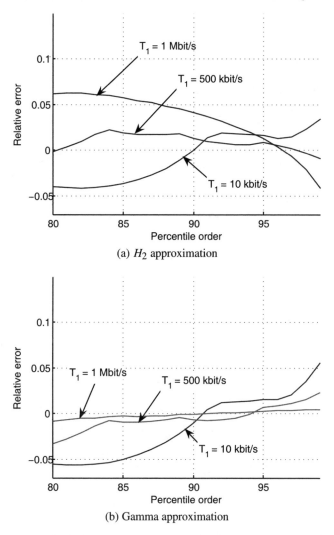

Figure 9.12   Relative error of actual and approximate delay percentiles for H_2 and gamma approximations for medium priority class.

Figures 9.12 and 9.13 show the relative error of the two delay DF approximations when compared to the delay DF obtained by simulation for medium and low priority classes, respectively. We can observe in both figures that for medium and low priority classes, and for both approximation approaches, the relative error is in the same order of magnitude as for the high priority shown in Figure 9.9. Furthermore, in both priority classes, both approximations obviously have an intersection with the simulated DF in all considered operating points. Fortunately, this intersection is close to the 95th percentile in the respective cases providing a very good approximation accuracy. The approximation accuracy is higher for the low priority class; overall, the gamma approximation is more accurate than the H_2 approximation.

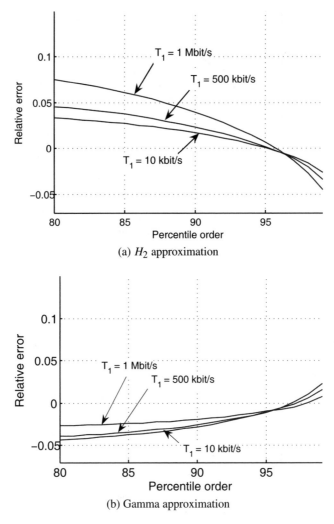

Figure 9.13 Relative error of actual and approximate delay percentiles for H_2 and gamma approximations for low priority class.

In particular, this applies to the 95th percentile as shown in Figure 9.14. In this figure, the relative error of the 95th delay percentile is plotted for the medium and low priority classes and for both approximations with varying offered traffic T_1 in the high priority class. Overall the approximation error is in the range between -2 and 3%. For both medium and low priority classes, the gamma approximation matches the 95th percentile better than the H_2 approximation method. The difference between the two approximation methods is bigger for the low priority class; for the medium priority class the approximations show a smaller difference. In general the approximation accuracy exhibits only a small dependency on the offered traffic of high priority class. There is no general tendency towards underestimation

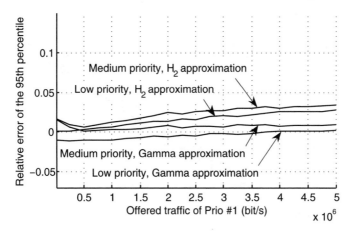

Figure 9.14 Relative error of the 95th percentile for H_2 and gamma approximations for medium and low priority classes with varying offered traffic in the high priority class.

or overestimation. While the H_2 approximation overestimates for both priority classes, the gamma approximation underestimates the low priority percentiles for low offered traffic of the high priority class, changes to slight overestimation for high offered traffic of the high priority class, and constantly overestimates for the medium priority class. The H_2 approximation shows a minimum in the error of the 95th percentile for a certain amount of high priority offered traffic, while the gamma approximation has a monotonous behavior.

9.6 Impact of Percentile Requirements on System Capacity

In this section we evaluate the impact of considering delay percentile requirements on the system capacity. We consider an example system with three priority classes, where a two-point packet size distribution with packet lengths being either 250 or 750 bytes is assumed for all priority classes. Unless otherwise stated, the offered traffic is 500 kbit/s in all priority classes. The mean delay requirements are chosen to depend on the classes in this example. The high priority packets are assumed to require the mean delay of 5 ms, the medium priority packets require 100 ms and the low priority packets need 1 s for the mean delay.

The required mean delay-bound system capacity, calculated with the method described in Section 4.3.5, is shown for all three priority classes in Figure 9.15(a). The offered traffic in the high priority class is varied between 10 kbit/s and 5 Mbit/s. Obviously, for low offered traffic in the high priority class, the mean delay-bound capacity is determined by the high priority class, while for higher offered traffic in the high priority class, the low priority traffic class requires the highest capacity.

Figure 9.15(b) shows the SCV in the waiting time that results for a system dimensioned according to the mean delay requirements of all priority classes, i.e. a system that is dimensioned so that for each value of high priority offered traffic the maximum of the three capacity requirement curves in Figure 9.15(a) is taken as the system capacity. For medium

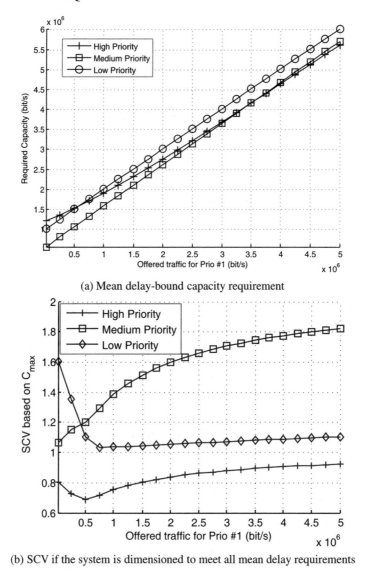

(a) Mean delay-bound capacity requirement

(b) SCV if the system is dimensioned to meet all mean delay requirements

Figure 9.15   Mean delay-bound capacity and the resultant SCV in the waiting time with varying the offered traffic of high priority class.

and low priority classes, the SCV is larger than one, indicating that the H_2 approximation method can be used for delay percentile bound dimensioning of the medium and low priority classes. From the development of SCV for the medium and low priority classes by varying the offered traffic of the high priority class, shown in Figure 9.15(b), we can conclude that a capacity increment under constant offered traffic and constant mean delay requirement leads to an increased SCV of the waiting time. We can thus expect that, for the mean delay-bound capacity calculation procedure, a switchover to the gamma approximation will not be required for the medium and low priority classes. The high priority class, however, has an SCV below one, suggesting that the gamma approximation must be used to approximate the waiting time DF for the high priority class.

In Figure 9.16, the delay percentile-bound capacities required by all three priority classes are shown for the 95th delay percentile requirement of 6.25 ms in the high priority class, 125 ms in the medium priority class, and 1.25 s in the low priority class. The upper envelope of the mean delay-bound capacity requirements shown in Figure 9.15(a) is taken as a starting point for the iterative calculation of the delay percentile-bound capacity requirements.

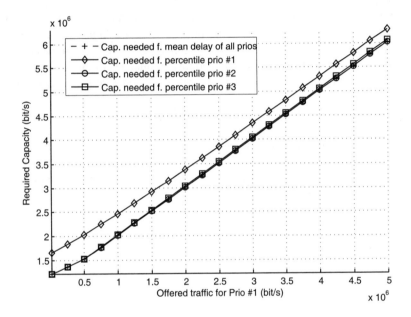

Figure 9.16 Capacity needed for the 95th delay percentile requirement of 6.25 ms, 125 ms and 1.25 s for the high, medium and low priority classes, respectively.

The high priority percentile requirement imposes significant additional capacity demand for the high offered traffic in the high priority class, whereas the mean delay requirements of the high priority class do not determine the mean delay-bound capacity. However, the additional amount of capacity needed to meet the delay percentile requirements decreases with increased traffic offered to the high priority class. The reason for this is that the upper envelope of the mean delay-bound capacity requirements increasingly over-fulfills the mean delay requirements on the high priority class, which reduces the variance of the delay distribution and therefore makes it easier to fulfill the delay percentile requirements.

The percentile specified for the medium priority class does not require additional capacity. Although not clearly visible in the figure, in the range of high traffic offered to the high priority class, the delay percentile criterion for the low priority class requires slightly more capacity than the mean delay requirement for the low priority class would need. This is caused by an increased variance in the delay distribution for the low priority class under increasing the offered traffic and system capacity for the high priority class; see Figure 9.11.

9.7   Conclusion

In this chapter, a method has been introduced to calculate the spectrum capacity requirements of IMT-Advanced systems so that delay percentiles specified for some services are met. It is shown by means of an example that the system capacity and therefore the required spectrum increase if not only the mean delay values but also the percentiles of delay are to be met.

Although the waiting time distribution for the M/G/1 queue under multi-modal service time distribution has discontinuities that are difficult to model with continuous functions, the proposed approximations for the waiting time DF for calculating the delay percentiles have been shown to be sufficiently accurate. The relative error is shown to be below 15% in the examples considered. The presented procedure can thus be considered to be robust and well suited for the purpose of spectrum requirement estimation according to delay percentile-based QoS requirements.

The sensitivity of the system capacity required by a priority class based on its delay percentile requirements depends on the variance of the delay, which in general is higher for lower priority classes. However, in realistic scenarios the low priority classes are likely to have less stringent or no delay percentile requirements. Then, percentile requirements of high priority classes may result in a significantly larger capacity required than needed under the mean delay constraints only.

10

Epilog: Result of WRC-07

Hitoshi Yoshino

International Telecommunication Union – Radiocommunication section (ITU-R) calculated the spectrum bandwidth requirement for both existing mobile cellular systems, including pre-IMT-2000, IMT-2000 and its enhancements (RATG 1), and IMT-Advanced (RATG 2) for the years 2010, 2015 and 2020. Figure 10.1 shows how the ITU-R concluded the spectrum requirement. The calculation was performed by using the spectrum estimation methodology described in Recommendation ITU-R M.1768 (Chapter 4 of this book) together with the market study parameter values in Report ITU-R M.2072 (Chapter 6), the radio-related parameter values in Report ITU-R M.2074 (Chapter 7), and other service category parameter values in Report ITU-R M.2078. The results are also included in Report ITU-R M.2078.

ITU-R summarized its technical studies as a Conference Preparatory Meeting (CPM) Report for the World Radiocommunication Conference 2007 (WRC-07) held in October–November 2007. At the WRC-07, additional spectrum bands for IMT were identified globally as well as for each ITU Region.

In this concluding chapter, we show the spectrum requirements predicted in Report ITU-R M.2078 and the spectrum requirements reported in the CPM Report. We then present the final spectrum identification for IMT as a result of WRC-07.

Report ITU-R M.2078

ITU-R has decided to take an approach to discuss a 'global common market', not markets on a Region-by-Region basis, since the mobile communication is inherently global due to its global roaming nature. However, ITU-R provided two spectrum bandwidth requirements for higher and lower user density settings, taking into account the different timings in the mobile market growth depending on the countries. The higher user density setting is used for those countries in which IMT-2000 has already been widely deployed and users have already had ample experiences with the broadband applications. The lower user density setting is prepared for those countries in which deployment of IMT-2000 has just been started. ITU-R sees that there are regional differences in the market development, i.e. in some parts

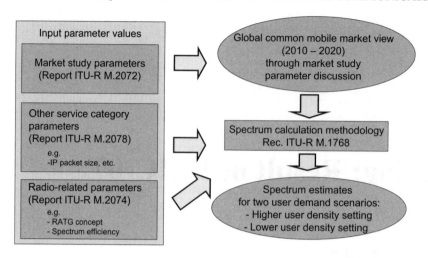

Figure 10.1 Concluding the spectrum estimation for IMT-Advanced.

of the world a particular level of market development may be reached earlier or later than in the average global common market.

Table 10.1 Predicted spectrum requirements (MHz) for RATG 1 and RATG 2.

User density	RATG 1			RATG 2			Total		
	2010	2015	2020	2010	2015	2020	2010	2015	2020
Lower	760	800	800	0	500	480	760	1300	1280
Higher	840	880	880	0	420	840	840	1300	1720

As indicated in Table 25 of Report ITU-R M.2078, Table 10.1 shows the predicted spectrum requirements for RATG 1 and RATG 2. We may note that for RATG 2 in the year 2015 the spectrum requirement in the lower user density setting (500 MHz) is larger than that in the higher user density setting (420 MHz). This is due to the rounding up procedure where the spectrum requirements are adjusted to become an integer multiple of the minimum spectrum deployment. Since the minimum spectrum is 120 MHz in this example case (pico cell), even a small difference in the initial spectrum requirements can end up in large differences in the spectrum requirements after adjustment. Also, the market parameter 'user density' and the radio-related parameter 'spectral efficiency' are different in the two user density settings. The lower user density setting assumes the spectral efficiency values lower than the higher user density setting does. For other years in Table 10.1, this anomaly does not appear.

Tables 10.2 and 10.3 (Table 26 of Report ITU-R M.2078) represent the spectrum requirements for lower and higher user density settings, respectively, for the various numbers of networks or network operators. When more than one network are deployed in a country, the total spectrum requirement may be higher in order to account for packaging the spectrum to integer multiples of 40 MHz for RATG 1.

Table 10.2 Predicted spectrum requirements (MHz) for various numbers of network with lower user density setting.

	Number of networks				
	1	2	3	4	5
RATG 1	800	880	840	1120	1000
RATG 2	480	560	720	800	1000
Total	1280	1440	1560	1920	2000

Table 10.3 Predicted spectrum requirements (MHz) for various numbers of network with higher user density setting.

	Number of networks				
	1	2	3	4	5
RATG 1	880	880	960	1120	1200
RATG 2	840	880	1020	1120	1300
Total	1720	1760	1980	2240	2500

Report ITU-R M.2078 predicted the total spectrum bandwidth requirement for RATG 1 and RATG 2 in the forecast year 2020 to be 1280 MHz in the lower user density setting and 1720 MHz in the higher user density setting, including the spectrum already in use or planned to be used for RATG 1. The Report concluded that additional spectrum is needed beyond that identified for IMT-2000 at the World Administrative Radiocommunication Conference 1992 (WARC-92) and the World Radiocommunication Conference 2000 (WRC-2000).

Conference Preparatory Meeting Report for WRC-07

ITU-R also developed a summary of technical studies within ITU-R and reported it to the WRC-07 as the Report of the CPM 'http://www.itu.int/md/R07-CPM-R-0001/en'. The CPM Report indicated the net additional spectrum requirement per ITU Region beyond that identified for IMT-2000 at both WARC-92 and WRC-2000, by taking into account that different amounts of spectrum have been identified for IMT-2000 in each Region as shown in Table 10.4.

The CPM listed candidate bands for IMT-2000 and IMT-Advanced (IMT as a whole) in order to accommodate the spectrum requirements predicted by ITU-R. It also described the advantages and disadvantages of each of these bands based on the results of ITU-R studies. The candidate bands were categorized into the following two sets:

Table 10.4   Spectrum requirements (MHz) reported to WRC-07 for the year 2020.[a]

User density setting	Predicted total spectrum	Region 1		Region 2		Region 3	
		Identified	Net additional	Identified	Net additional	Identified	Net additional
Lower	1280	693	587	723	557	749	531
Higher	1720	693	1027	723	997	749	971

[a]Prediction based on one network deployment.

- bands for area coverage, i.e. for connecting the unconnected in ITU's vision: 410–430 MHz, 450–470 MHz and 470–806/862 MHz;

- bands for high-capacity communications, i.e. for broadband mobile access: 2.3–2.4 GHz, 2.7–2.9 GHz, 3.4–4.2 GHz and 4.4–4.99 GHz.

The CPM Report also mentioned that due account should be taken of the radio communication services to which the frequency bands were currently allocated. The spectrum-sharing studies with other radio services in the candidate bands were very important in order to obtain a green light of the spectrum identification from ITU-R. The summary of the results of these spectrum-sharing studies were also included in the CPM Report based on the ITU-R Reports such as:

- sharing with other radiocommunication services in the 450–470 MHz band (Report ITU-R M.2110);

- sharing with airport surveillance radars and meteorological radars in the 2700–2900 MHz band (Report ITU-R M.2112);

- sharing with geostationary satellite networks in the fixed satellite services in the 3400–4200 MHz and 4500–4800 MHz bands (Report ITU-R M.2109).

Result of WRC-07

ITU-R World Radiocommunication Conference 2007 (WRC-07) was held from 22 October to 16 November 2007, in Geneva, Switzerland. The successful development of the new spectrum calculation methodology and subsequent estimation of the spectrum requirement for IMT with the methodology gave confidence to WRC-07 in terms of the future need of IMT spectrum.

The WRC-07 started its discussion based on the spectrum requirements for the year 2020 which were predicted in the ITU-R study. Since the ITU-R study showed that additional spectra were required for IMT, there was no particular issue with regard to the conducted requirement studies themselves. The WRC-07 was therefore able to focus its discussion on which bands could be identified for the use of IMT, taking into account the need for existing radio services in the candidate bands and the possibility of sharing spectra with these existing services.

The WRC-07 extensively discussed the availability of new spectra for IMT by considering the balance between the new spectrum need for IMT and the need for incumbent radio

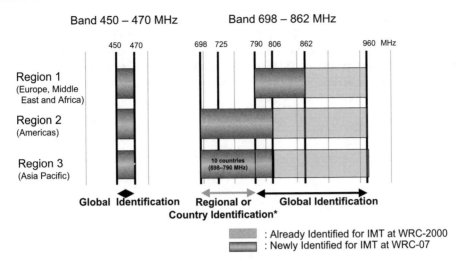

Figure 10.2 The bands below 1 GHz for area coverage.

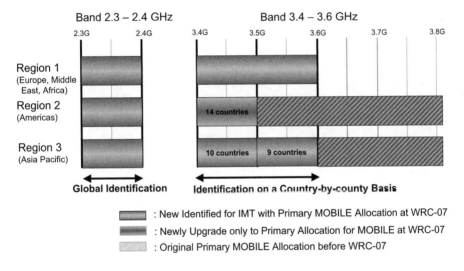

Figure 10.3 The bands above 1 GHz for high-capacity communications.

services such as fixed satellite service (FSS) and broadcasting service (BS). The WRC-07 finally decided to identify the spectra ranging from 120 MHz up to 392 MHz, 228 MHz and 428 MHz in ITU-R Region 1 (Europe, Middle East and Africa), Region 2 (Americas), and Region 3 (Asia Pacific), respectively, for the use of IMT, as shown in Figures 10.2 and 10.3. It is noteworthy that the amounts are different in different countries.

The new identification at WRC-07 may partially fulfill the spectrum requirements before the year 2015 according to the ITU-R study. Spectrum sharing with other existing services will be another key issue to the success in obtaining a new spectrum for IMT, and ultimately for sustainable development of IMT services. Further ITU-R study will be required in preparation for future WRCs.

Appendices

Appendices

Appendix A

Derivation of Formulas by Queueing Theory

Hideaki Takagi

In this appendix, we derive the basic formulas used in the methodology for determining the capacity requirement as shown in Table A.1. These formulas are derived by the theory of queues. Erlang-B formula for the blocking probability in a loss system, Erlang-C formula for the wait probability in a delay system, and Cobham's formula for the average waiting time in an M/G/1 nonpreemptive priority queue appear in introductory textbooks of queueing theory such as Cooper (1991), Gross and Harris (1998), Kleinrock (1975), and Kleinrock (1976) as well as in monographs on teletraffic engineering such as Fujiki and Gambe (1980) and Syski (1986). We refer to Wolff (1989) for the proofs of fundamental properties of queues. The multidimensional Erlang-B formula for the blocking probability in a loss system with multiple classes of calls and multiple server occupation is an example of so-called product-form solution to a network of queues that can be modeled by a reversible Markov process (Iversen). However, we present its analysis and computational algorithm in an elementary manner by following Kaufman (1981).

A basic queueing model consists of an arrival stream of customers, a waiting room and a set of servers. Various queueing models are conveniently described by the so-called *Kendall's notation* 'A/B/s/K' as explained in Gross and Harris (1998, p. 8) and Kleinrock (1975, p. 399). In this notation, the first parameter 'A' indicates the arrival process such as 'M' for the Poisson process. The second parameter 'B' indicates the service time distribution such as 'M' for the exponential distribution, 'D' for the deterministic, i.e. constant service time, and 'G' for the general distribution. The third parameter s denotes the number of servers. The fourth parameter K denotes the maximum number of customers that can be accommodated in the whole system, i.e. both in the waiting room and servers. The fourth parameter may be omitted if there is no restriction on the capacity of the waiting room; this model is a delay system. If $K = s$, the model is a loss system, because there is no waiting room. The first three parameters seem to be quite standard. However, the reader should note that the fourth parameter is used with different meaning in some books, e.g. Walke (2002, p. 1043).

Table A.1   Sections where formulas of queueing theory are used.

Formula	System	Traffic	Section
Erlang-B formula for a loss system	IMT-2000	circuit-switched	3.3.2
Erlang-C formula for a delay system	IMT-2000	packet-switched	3.3.2
Multidimensional Erlang-B formula	IMT-Advanced	circuit-switched	4.3.4
M/G/1 nonpreemptive priority queue	IMT-Advanced	packet-switched	4.3.5

A.1   Erlang-B Formula for a Loss System

Let us consider an M/M/s/s loss system depicted in Figure A.1. There are s servers and no waiting room. Calls arrive in a Poisson process with rate λ. The service time of each call has exponential distribution with mean $1/\mu$. Calls that arrive when all servers are busy are blocked and lost.

We define the state of the system by the number of calls present in the system. The state space is finite. The state follows a birth-and-death process, for which the state transition rate diagram is shown in Figure A.2.

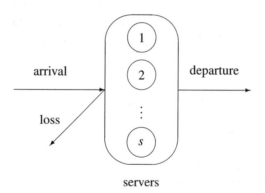

Figure A.1   A loss system.

Let p_k be the probability that there are k calls in the system at an arbitrary time in equilibrium, where $k = 0, 1, 2, \ldots, s$. Then the set of balance equations for $\{p_k; 0 \le k \le s\}$ is given by

$$\mu p_1 = \lambda p_0$$

$$\lambda p_{k-1} + (k+1)\mu p_{k+1} = (\lambda + k\mu)p_k, \quad 1 \le k \le s - 1 \qquad (A.1)$$

$$\lambda p_{s-1} = s\mu p_s.$$

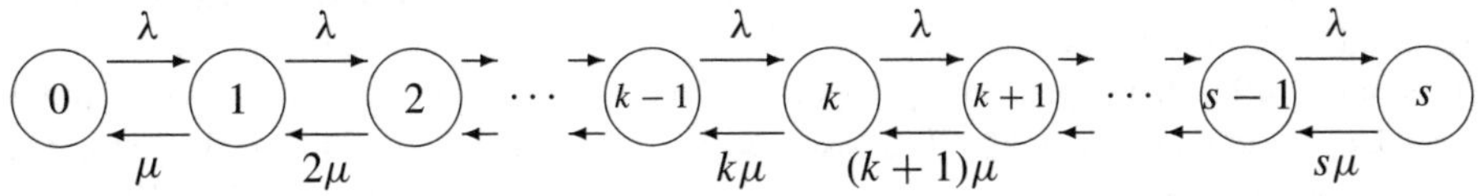

Figure A.2 State transition rate diagram for an M/M/s/s loss system.

This is equivalent to the set

$$k\mu p_k = \lambda p_{k-1}, \quad 1 \leq k \leq s \tag{A.2}$$

which gives

$$p_k = \frac{\lambda}{k\mu} p_{k-1} = \frac{\lambda^2}{k(k-1)\mu^2} p_{k-2} = \cdots = \frac{1}{k!}\left(\frac{\lambda}{\mu}\right)^k p_0, \quad 1 \leq k \leq s. \tag{A.3}$$

Writing $a := \lambda/\mu$ for the offered traffic, and finding p_0 from the normalization condition

$$\sum_{k=0}^{s} p_k = 1, \tag{A.4}$$

we obtain the probability distribution for the number of calls in the system

$$p_k = p_0 \frac{a^k}{k!} = \frac{a^k/k!}{\sum_{j=0}^{s}(a^j/j!)}, \quad 0 \leq k \leq s. \tag{A.5}$$

This is the *truncated Poisson distribution*.

Calls that arrive when all servers are used are blocked. From the *PASTA* (Poisson arrivals see time averages) property (Wolff 1989, pp. 293–297), the probability that there are k calls in the system immediately before an arrival time equals the probability that there are k calls in the system at an arbitrary time given in Equation (A.5). Therefore, the *blocking probability* is given by

$$p_s = \frac{u^s/s!}{\sum_{j=0}^{s}(a^j/j!)} := E_B(s, a). \tag{A.6}$$

This is called the *Erlang-B formula* or *Erlang's loss formula*. This formula was first derived by A. K. Erlang in 1917.[1]. This result is used in the methodology of required capacity calculation for the circuit-switched services in IMT-2000 systems in Section 3.3.2.

We note the relation

$$E_B(0, a) = 1; \quad E_B(s, a) = \frac{a E_B(s-1, a)}{s + a E_B(s-1, a)}, \quad s \geq 1 \tag{A.7}$$

which was very useful to calculate $E_B(s, a)$ for $s = 1, 2, \ldots$, in this order recursively for a given value of a in the era of B. C. (before computer).

---

[1]See Brockmeyer *et al.* (1948) for the life and works of A. K. Erlang.

Erlang-B formula is derived in Cooper (1991, p. 80), Fujiki and Gambe (1980, p. 47), Gross and Harris (1998, p. 80), Kleinrock (1975, p. 106), and Syski (1986, p. 147).

It is noteworthy that Equations (A.5) and (A.6) are valid even when the service time has general distribution, i.e. for an M/G/s/s loss system. In such a case, $1/\mu$ in the above formulas is replaced by the average service time. This property is called the *insensitivity* (Wolff 1989, pp. 271–273).

A.2   Erlang-C Formula for a Delay System

Let us consider an M/M/s delay system depicted in Figure A.3. There are s servers and a waiting room of infinite capacity. Calls arrive in a Poisson process with rate λ. The service time of each call has exponential distribution with mean $1/\mu$. Calls that arrive when all servers are busy wait in the waiting room.

We define the state of the system by the number of calls present in the system. The state space is infinite. The state follows a birth-and-death process, for which the state transition rate diagram is shown in Figure A.4.

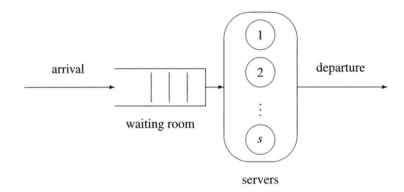

Figure A.3   A delay system.

Let p_k be the probability that there are k calls in the system at an arbitrary time in equilibrium, where $k = 0, 1, 2, \ldots$. Then the set of balance equations for $\{p_k; k \geq 0\}$ is given by

$$\mu p_1 = \lambda p_0$$
$$\lambda p_{k-1} + (k+1)\mu p_{k+1} = (\lambda + k\mu)p_k, \quad 1 \leq k \leq s-1$$
$$\lambda p_{k-1} + s\mu p_{k+1} = (\lambda + s\mu)p_k, \quad k \geq s. \tag{A.8}$$

This is equivalent to the set

$$k\mu p_k = \lambda p_{k-1}, \quad 1 \leq k \leq s; \quad s\mu p_k = \lambda p_{k-1}, \quad k \geq s \tag{A.9}$$

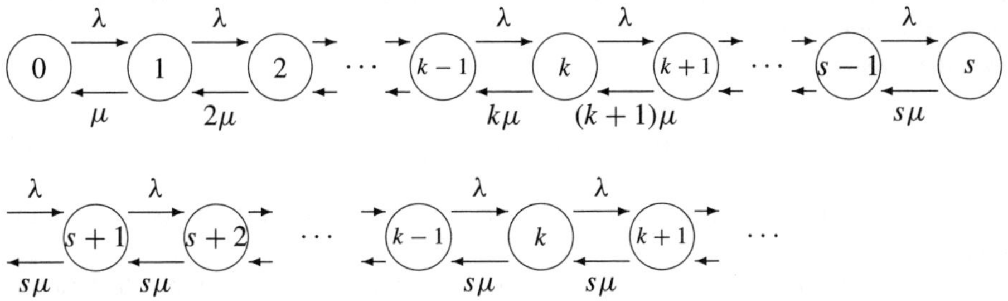

Figure A.4 State transition rate diagram for an M/M/s delay system.

which gives

$$p_k = \frac{\lambda}{k\mu}p_{k-1} = \frac{\lambda^2}{k(k-1)\mu^2}p_{k-2} = \cdots = \frac{1}{k!}\left(\frac{\lambda}{\mu}\right)^k p_0, \quad 1 \le k \le s \quad (A.10)$$

$$p_k = \frac{\lambda}{s\mu}p_{k-1} = \left(\frac{\lambda}{s\mu}\right)^2 p_{k-2} = \cdots = \left(\frac{\lambda}{s\mu}\right)^{k-s} p_s$$

$$= \frac{(\lambda/\mu)^k}{s!s^{k-s}}p_0 = \frac{s^s}{s!}\left(\frac{\lambda}{s\mu}\right)^k p_0, \quad k \ge s. \quad (A.11)$$

Writing $a := \lambda/\mu$ for the offered traffic, we determine p_0 from the normalization condition

$$\sum_{k=0}^{\infty} p_k = 1. \quad (A.12)$$

When $a/s < 1$ (the stability condition), we obtain

$$\frac{1}{p_0} = \sum_{k=0}^{s-1}\frac{a^k}{k!} + \frac{s^s}{s!}\sum_{k=s}^{\infty}\left(\frac{a}{s}\right)^k = \sum_{k=0}^{s-1}\frac{a^k}{k!} + \frac{a^s}{(s-1)!(s-a)}. \quad (A.13)$$

Thus we obtain the probability distribution for the number of calls in the system

$$p_k = \begin{cases} p_0\dfrac{a^k}{k!}, & 0 \le k \le s \\[2mm] p_0\dfrac{s^s}{s!}\left(\dfrac{a}{s}\right)^k = p_s\left(\dfrac{a}{s}\right)^{k-s}, & k \ge s. \end{cases} \quad (A.14)$$

Calls that arrive when all servers are used must wait in the waiting room. Owing to the PASTA property, the wait probability is given by

$$P\{W > 0\} = \sum_{k=s}^{\infty} p_k = p_s \sum_{k=s}^{\infty}\left(\frac{a}{s}\right)^{k-s} = \frac{p_s s}{s-a} = \frac{p_0 a^s}{(s-1)!(s-a)}$$

$$= \frac{a^s/(s-1)!(s-a)}{\sum_{j=0}^{s-1} a^j/j! + a^s/(s-1)!(s-a)} := E_C(s, a). \quad (A.15)$$

This is called the *Erlang-C formula* or *Erlang's delay formula*. This formula was also derived by A. K. Erlang in 1917.

We may note the relationship

$$E_C(s, a) = \frac{s E_B(s, a)}{s - a + a E_B(s, a)} \tag{A.16}$$

where $E_B(s, a)$ is the blocking probability in the corresponding M/M/s/s loss system given in Equation (A.6).

We now find the distribution function for the waiting time W by assuming the first-come first-served (FCFS) discipline. Suppose that no servers are idle during a time interval x. Then, the number of service completions during x follows the Poisson distribution with mean $s\mu x$. Let N^- be the number of calls present in the system immediately before an arrival. Hence

$$P\{W > x \mid N^- = k\} = P\{\text{Number of service completions before } x \le k - s\}$$

$$= \sum_{j=0}^{k-s} \frac{(s\mu x)^j}{j!} e^{-s\mu x}, \quad x \ge 0. \tag{A.17}$$

It follows from PASTA again and the theorem of total probability that

$$P\{W > x\} = \sum_{k=s}^{\infty} p_k \cdot P\{W > x \mid N^- = k\}$$

$$= \sum_{k=s}^{\infty} p_s \left(\frac{a}{s}\right)^{k-s} \sum_{j=0}^{k-s} \frac{(s\mu x)^j}{j!} e^{-s\mu x}$$

$$= p_s \, e^{-s\mu x} \sum_{j=0}^{\infty} \frac{(s\mu x)^j}{j!} \sum_{k=s+j}^{\infty} \left(\frac{a}{s}\right)^{k-s}$$

$$= p_s \, e^{-s\mu x} \sum_{j=0}^{\infty} \frac{(s\mu x)^j}{j!} \cdot \left(\frac{a}{s}\right)^j \frac{1}{1 - a/s}$$

$$= \frac{p_s s}{s - a} e^{-s\mu x} \sum_{j=0}^{\infty} \frac{(a\mu x)^j}{j!}$$

$$= E_C(s, a) \, e^{-(s-a)\mu x}, \quad x \ge 0. \tag{A.18}$$

This result is used in the methodology of required capacity calculation for the packet-switched services in IMT-2000 systems in Section 3.3.2.

Though not used in the methodology, it is interesting to note that the distribution function for the waiting time of those calls that are forced to wait upon arrival is given by

$$P\{W > x \mid W > 0\} = \frac{P\{W > x, W > 0\}}{P\{W > 0\}} = \frac{P\{W > x\}}{P\{W > 0\}}$$

$$= e^{-(s-a)\mu x}, \quad x > 0, \tag{A.19}$$

which is an exponential distribution with average $1/[(s - a)\mu]$.

We also note that the time average of the number of calls present in the waiting room, denoted by L, at an arbitrary time is given by

$$E[L] = \sum_{k=s+1}^{\infty} (k-s)p_k = p_s \sum_{k=s+1}^{\infty} (k-s) \left(\frac{a}{s}\right)^{k-s} = \frac{p_s a s}{(s-a)^2}. \tag{A.20}$$

The waiting time averaged over the calls is given by

$$E[W] = \int_0^{\infty} P\{W > x\} \, dx = \frac{E_C(s,a)}{(s-a)\mu} = \frac{p_s s}{(s-a)^2 \mu}. \tag{A.21}$$

Hence we can confirm the relation

$$E[L] = \lambda E[W]. \tag{A.22}$$

This is an instance of *Little's law* (Wolff 1989, pp. 235–238) applied to the calls in the waiting room.

Erlang-C formula and the waiting time distribution are derived in Cooper (1991, pp. 90–98), Fujiki and Gambe (1980, p. 75), Gross and Harris (1998, pp. 69–73), Kleinrock (1975, p. 103), and Syski (1986, pp. 238–239).

A.3 Multidimensional Erlang-B Formula

An M/M/s/s loss system with N classes of calls and multiple server occupation is used in the methodology of required capacity calculation for the circuit-switched service categories in IMT-Advanced systems in Section 4.3.4. Before presenting its analysis and computational algorithm which may look formidable at first glance, we study a simple case with two classes of calls and single server occupation. The two-dimensional state transition rate diagram in Figure A.5 helps the reader understand the analysis of the latter system.

A.3.1 Two classes of calls with single server occupation

We first consider an M/M/s/s loss system with two classes of calls. There are s servers and no waiting room. Calls of class 1 and class 2 arrive in independent Poisson processes with rate λ_1 and with rate λ_2, respectively. Each call occupies one server. The service time of each call has exponential distribution with average $1/\mu_1$ for class 1 and with average $1/\mu_2$ for class 2. Any call that arrives when all servers are busy is blocked.

We define the state of the system by a combination of the number of calls of class 1 and the number of calls of class 2 that are present in the system. It is denoted by (j, k), where j and k are the numbers of calls of class 1 and class 2, respectively, present in the system. The state space is finite

$$\Omega := \{(j, k) : j + k \leq s, j \geq 0, k \geq 0\}. \tag{A.23}$$

The number of states is $(s+1)(s+2)/2$. The state follows a two-dimensional birth-and-death process, for which the state transition rate diagram is shown in Figure A.5.

Let $p_{j,k}$ be the probability that there are j calls of class 1 and k calls of class 2 present in the system at an arbitrary time in equilibrium. Referring to Figure A.6(a), the set of *global*

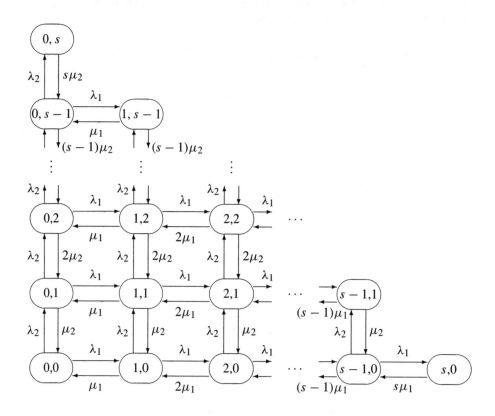

Figure A.5 State transition rate diagram for an M/M/s/s loss system with two classes.

balance equations for $\{p_{j,k}; (j,k) \in \Omega\}$ is given by

$$\mu_1 p_{1,0} + \mu_2 p_{0,1} = (\lambda_1 + \lambda_2) p_{0,0}$$

$$\lambda_1 p_{j-1,0} + (j+1)\mu_1 p_{j+1,0} + \mu_2 p_{j,1} = (\lambda_1 + \lambda_2 + j\mu_1) p_{j,0}, \quad 1 \leq j \leq s-1$$

$$\lambda_1 p_{s-1,0} = s\mu_1 p_{s,0}$$

$$\lambda_2 p_{0,k-1} + \mu_1 p_{1,k} + (k+1)\mu_2 p_{0,k+1} = (\lambda_1 + \lambda_2 + k\mu_2) p_{0,k}, \quad 1 \leq k \leq s-1$$

$$\lambda_2 p_{0,s-1} = s\mu_2 p_{0,s} \qquad\qquad (A.24)$$

$$\lambda_1 p_{j-1,k} + \lambda_2 p_{j,k-1} + (j+1)\mu_1 p_{j+1,k} + (k+1)\mu_2 p_{j,k+1}$$

$$= (\lambda_1 + \lambda_2 + j\mu_1 + k\mu_2) p_{j,k}, \quad 1 \leq j,k \leq s-1, j+k \leq s-1$$

$$\lambda_1 p_{j-1,k} + \lambda_2 p_{j,k-1} = (j\mu_1 + k\mu_2) p_{j,k}, \quad 1 \leq j,k \leq s-1, j+k = s.$$

The number of equations is $(s+1)(s+2)/2$, which is the same as the number of states. One of the equations is redundant as they are homogeneous. Another equation to determine the

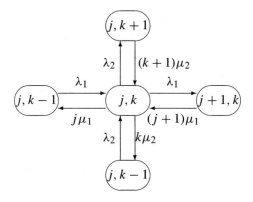

(a) Global balance

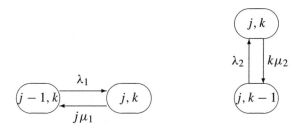

(b) Local balance

Figure A.6 Global and local balance for an M/M/s/s loss system with two classes.

set $\{p_{j,k}; (j, k) \in \Omega\}$ uniquely is given by the normalization condition in Equation (A.27) below.

The set of equations in (A.24) is equivalent to the following set of *local balance equations*:

$$j\mu_1 p_{j,k} = \lambda_1 p_{j-1,k}, \quad 1 \le j \le s, 0 \le k \le s$$

$$k\mu_2 p_{j,k} = \lambda_2 p_{j,k-1}, \quad 0 \le j \le s, 1 \le k \le s. \tag{A.25}$$

See Figure A.6(b). These equations are satisfied by the *product-form solution*:

$$p_{j,k} = \frac{1}{G(s)} \frac{(a_1)^j}{j!} \cdot \frac{(a_2)^k}{k!}, \quad (j, k) \in \Omega, \tag{A.26}$$

where $a_1 := \lambda_1/\mu_1$ is the offered traffic of class 1 and $a_2 := \lambda_2/\mu_2$ is the offered traffic of class 2. The constant $G(s)$ can be found from the normalization condition

$$\sum_{(j,k)\in\Omega} p_{j,k} = 1 \tag{A.27}$$

as follows:

$$G(s) = \sum_{(j,k)\in\Omega} \frac{(a_1)^j}{j!} \cdot \frac{(a_2)^k}{k!} = \sum_{j=0}^{s}\sum_{k=0}^{s-j} \frac{(a_1)^j}{j!} \cdot \frac{(a_2)^k}{k!}$$

$$= \sum_{i=0}^{s}\sum_{j=0}^{i} \frac{(a_1)^j}{j!} \cdot \frac{(a_2)^{i-j}}{(i-j)!} = \sum_{i=0}^{s} \frac{(a_2)^i}{i!} \sum_{j=0}^{i} \binom{i}{j}\left(\frac{a_1}{a_2}\right)^j$$

$$= \sum_{i=0}^{s} \frac{(a_2)^i}{i!}\left(1 + \frac{a_1}{a_2}\right)^i = \sum_{i=0}^{s} \frac{(a_1 + a_2)^i}{i!}, \tag{A.28}$$

where we have changed the variable by $j + k = i$ and used the binomial theorem. Hence we obtain the state probability explicitly

$$p_{j,k} = \frac{(a_1)^j/j! \cdot (a_2)^k/k!}{\sum_{i=0}^{s}(a_1 + a_2)^i/i!}, \quad (j,k) \in \Omega. \tag{A.29}$$

Let Ω_l be the state space in which the total number of calls in the system is exactly l:

$$\Omega_l := \{(j,k) : j + k = l, j \geq 0, k \geq 0\}, \quad 0 \leq l \leq s. \tag{A.30}$$

Then we have the probability that there are l calls in the system

$$P\{(j,k) \in \Omega_l\} = \sum_{(j,k)\in\Omega_l} p_{j,k} = \sum_{j=0}^{l} p_{j,l-j} = \frac{1}{G(s)}\sum_{j=0}^{l} \frac{(a_1)^j}{j!} \cdot \frac{(a_2)^{l-j}}{(l-j)!}$$

$$= \frac{1}{G(s)}\frac{(a_1 + a_2)^l}{l!} = \frac{(a_1 + a_2)^l}{l!} \bigg/ \sum_{i=0}^{s} \frac{(a_1 + a_2)^i}{i!}, \quad 0 \leq l \leq s. \tag{A.31}$$

This is a truncated Poisson distribution just like Equation (A.5) for an M/M/s/s loss system with a single class of offered traffic $a = a_1 + a_2$.

An arriving call of either class is blocked if all servers are used when it arrives. This happens when the system is in state Ω_s when a call arrives. Due to the PASTA property, the blocking probability is given by

$$P^B = P\{(j,k) \in \Omega_s\} = \frac{(a_1 + a_2)^s}{s!} \bigg/ \sum_{i=0}^{s} \frac{(a_1 + a_2)^i}{i!}. \tag{A.32}$$

Since

$$\frac{(a_1 + a_2)^s}{s!} = \sum_{i=0}^{s} \frac{(a_1 + a_2)^i}{i!} - \sum_{i=0}^{s-1} \frac{(a_1 + a_2)^i}{i!}, \tag{A.33}$$

the blocking probability can be expressed as

$$P^B = 1 - \frac{G(s-1)}{G(s)}. \tag{A.34}$$

A.3.2 Several classes of calls with multiple server occupation

We now consider an M/M/s/s loss system with N classes of calls and multiple server occupation. There are s servers and no waiting room. Calls of class n arrive in a Poisson process with rate λ_n, independent of the arrival processes of other classes ($1 \leq n \leq N$). It is further assumed that each call of class n occupies m_n servers simultaneously during the service ($1 \leq n \leq N$). Let

$$\boldsymbol{m} \equiv (m_1, m_2, \ldots, m_N).$$

The service time of each call of class n has exponential distribution with average $1/\mu_n$ ($1 \leq n \leq N$). When the service of a call is finished, all servers occupied by the call are released immediately. Any call that arrives when all servers are busy is blocked.

We follow Kaufman (1981) for the analysis and the computational algorithm of the blocking probability. The state of the system is denoted by

$$\boldsymbol{k} \equiv (k_1, k_2, \ldots, k_N) \tag{A.35}$$

where k_n is the number of calls of class n that are present in the system ($1 \leq n \leq N$). The state space is finite

$$\Omega := \{\boldsymbol{k} : \boldsymbol{m} \cdot \boldsymbol{k} \leq s, k_n \geq 0 \ (1 \leq n \leq N)\}, \tag{A.36}$$

where

$$\boldsymbol{m} \cdot \boldsymbol{k} \equiv \sum_{n=1}^{N} m_n k_n. \tag{A.37}$$

In order to write the global balance equations, we introduce the notation

$$\boldsymbol{k}_n^+ \equiv (k_1, \ldots, k_{n-1}, k_n + 1, k_{n+1}, \ldots, k_N)$$
$$\boldsymbol{k}_n^- \equiv (k_1, \ldots, k_{n-1}, k_n - 1, k_{n+1}, \ldots, k_N) \tag{A.38}$$

and

$$\delta_n^+(\boldsymbol{k}) := \begin{cases} 1, & \boldsymbol{k}_n^+ \in \Omega \\ 0, & \text{otherwise;} \end{cases} \qquad \delta_n^-(\boldsymbol{k}) := \begin{cases} 1, & \boldsymbol{k}_n^- \in \Omega \\ 0, & \text{otherwise.} \end{cases} \tag{A.39}$$

The set of global balance equations is given by

$$\sum_{n=1}^{N} \lambda_n \delta_n^-(\boldsymbol{k}) P(\boldsymbol{k}_n^-) + \sum_{n=1}^{N} (k_n + 1) \mu_n \delta_n^+(\boldsymbol{k}) P(\boldsymbol{k}_n^+)$$
$$= \left[\sum_{n=1}^{N} \lambda_n \delta_n^+(\boldsymbol{k}) + \sum_{n=1}^{N} k_n \mu_n \delta_n^-(\boldsymbol{k}) \right] P(\boldsymbol{k}), \quad \boldsymbol{k} \in \Omega. \tag{A.40}$$

This is equivalent to the following set of local balance equations

$$k_n \mu_n \delta_n^-(\boldsymbol{k}) P(\boldsymbol{k}) = \lambda_n \delta_n^-(\boldsymbol{k}) P(\boldsymbol{k}_n^-), \quad 1 \leq n \leq N, \boldsymbol{k} \in \Omega. \tag{A.41}$$

The solution is given in the product form:

$$P(\boldsymbol{k}) = \frac{1}{G(\Omega)} \prod_{n=1}^{N} \frac{(a_n)^{k_n}}{k_n!}, \quad \boldsymbol{k} \in \Omega \tag{A.42}$$

with

$$G(\Omega) = G(s) = \sum_{k \in \Omega} \prod_{n=1}^{N} \frac{(a_n)^{k_n}}{k_n!} \tag{A.43}$$

where $a_n := \lambda_n/\mu_n$ is the offered traffic of class n ($1 \leq n \leq N$). It can be easily confirmed by substitution that the solution in (A.42) satisfies Equations (A.41) because

$$P(\boldsymbol{k}_n^-) = \frac{(a_1)^{k_1}}{k_1!} \cdots \frac{(a_{n-1})^{k_{n-1}}}{k_{n-1}!} \cdot \frac{(a_n)^{k_n-1}}{(k_n-1)!} \cdot \frac{(a_{n+1})^{k_{n+1}}}{k_{n+1}!} \cdots \frac{(a_N)^{k_N}}{k_N!}. \tag{A.44}$$

An arriving call of either class is blocked if all servers are used when it arrives. The state space in which an arriving call of class n is blocked is given by

$$B_n := \{\boldsymbol{k} \in \Omega : \boldsymbol{k}_n^+ \notin \Omega\} = \Omega \setminus \{\boldsymbol{k} \in \Omega : 0 \leq \boldsymbol{m} \cdot \boldsymbol{k} \leq s - m_n\}. \tag{A.45}$$

Therefore, the blocking probability for calls of class n is given by

$$P_n^{\mathrm{B}}(s) = \sum_{k \in B_n} P(\boldsymbol{k}) = \frac{G(B_n)}{G(\Omega)} \tag{A.46}$$

where

$$G(B_n) = \sum_{k \in B_n} \prod_{i=1}^{N} \frac{(a_i)^{k_i}}{k_i!} = G(s) - G(s - m_n). \tag{A.47}$$

Thus we obtain

$$P_n^{\mathrm{B}}(s) = 1 - \frac{G(s - m_n)}{G(s)}. \tag{A.48}$$

We proceed to the computational algorithm. We define

$$q(j) := P\{\boldsymbol{m} \cdot \boldsymbol{k} = j\} = \sum_{k \in \Omega_j} P(\boldsymbol{k}), \quad 0 \leq j \leq s \tag{A.49}$$

where Ω_j is the state space in which j servers are occupied:

$$\Omega_j := \{\boldsymbol{k} \in \Omega : \boldsymbol{m} \cdot \boldsymbol{k} = j, k_n \geq 0 \ (1 \leq n \leq N)\}, \quad 0 \leq j \leq s. \tag{A.50}$$

It is clear that the set $\{\Omega_j; 0 \leq j \leq s\}$ is a partition of Ω:

$$\Omega_i \cap \Omega_j = \emptyset \quad i \neq j; \quad \Omega = \Omega_0 \cup \Omega_1 \cup \Omega_2 \cup \cdots \cup \Omega_s. \tag{A.51}$$

It follows that

$$\sum_{j=0}^{s} q(j) = \sum_{j=0}^{s} \sum_{k \in \Omega_j} P(\boldsymbol{k}) = \sum_{k \in \Omega} P(\boldsymbol{k}) = 1. \tag{A.52}$$

Then we have

$$G(j) = G(s) \sum_{l=0}^{j} q(l), \quad 0 \leq j \leq s. \tag{A.53}$$

Conversely, since $1 = G(0) = G(s)q(0)$ we obtain

$$q(0) = \frac{1}{G(s)}; \quad q(j) = \frac{G(j) - G(j-1)}{G(s)}, \quad 1 \leq j \leq s. \tag{A.54}$$

The key for the algorithm is the following lemma.

Lemma

$$a_n q(j - m_n) = q(j) E[k_n \mid \boldsymbol{m} \cdot \boldsymbol{k} = j], \quad 1 \le n \le N, \ 0 \le j \le s. \tag{A.55}$$

Proof. The local balance equations in (A.41) can be written as

$$a_n \delta_n(\boldsymbol{k}) P(\boldsymbol{k}_n^-) = k_n P(\boldsymbol{k}), \quad 1 \le n \le N \tag{A.56}$$

where

$$\delta_n(\boldsymbol{k}) := \begin{cases} 1, & k_n \ge 1 \\ 0, & \text{otherwise.} \end{cases} \tag{A.57}$$

Summing both sides of (A.56) over Ω_j, we obtain

$$a_n \sum_{\boldsymbol{k} \in \Omega_j} \delta_n(\boldsymbol{k}) P(\boldsymbol{k}_n^-) = \sum_{\boldsymbol{k} \in \Omega_j} k_n P(\boldsymbol{k}). \tag{A.58}$$

We calculate each side of this equation. First, by noting that

$$P(\boldsymbol{k} \mid \boldsymbol{m} \cdot \boldsymbol{k} = j) = \begin{cases} P(\boldsymbol{k})/q(j), & \boldsymbol{k} \in \Omega_j \\ 0, & \text{otherwise} \end{cases} \tag{A.59}$$

the right-hand side of (A.58) becomes

$$\sum_{\boldsymbol{k} \in \Omega_j} k_n P(\boldsymbol{k}) = \sum_{\boldsymbol{k} \in \Omega_j} k_n \frac{P(\boldsymbol{k})}{q(j)} \cdot q(j)$$

$$= q(j) \sum_{\boldsymbol{k} \in \Omega_j} k_n P(\boldsymbol{k} \mid \boldsymbol{m} \cdot \boldsymbol{k} = j)$$

$$= q(j) E[k_n \mid \boldsymbol{m} \cdot \boldsymbol{k} = j]. \tag{A.60}$$

Next, on the left-hand side of (A.58), we have

$$\sum_{\boldsymbol{k} \in \Omega_j} \delta_n(\boldsymbol{k}) P(\boldsymbol{k}_n^-) = \sum_{\boldsymbol{k} \in \Omega_j \cap \{k_n \ge 1\}} P(\boldsymbol{k}_n^-). \tag{A.61}$$

However, since

$$\boldsymbol{m} \cdot \boldsymbol{k} = \sum_{l=1}^{N} m_l k_l = \sum_{l \ne n}^{N} m_l k_l + m_n(k_n - 1) + m_n = \boldsymbol{m} \cdot \boldsymbol{k}_n^- + m_n, \tag{A.62}$$

it follows that

$$\Omega_j \cap \{k_n \ge 1\} = \{\boldsymbol{k} : \boldsymbol{m} \cdot \boldsymbol{k} = j, \ k_n \ge 1, \ k_l \ge 0 \ (l \ne n)\}$$

$$= \{\boldsymbol{k}_n^- : \boldsymbol{m} \cdot \boldsymbol{k}_n^- = j - m_n, \ (\boldsymbol{k}_n^-)_l \ge 0 \ (1 \le l \le N)\}. \tag{A.63}$$

Hence we obtain

$$\sum_{\boldsymbol{k} \in \Omega_j} \delta_n(\boldsymbol{k}) P(\boldsymbol{k}_n^-) = q(j - m_n). \tag{A.64}$$

q. e. d.

This lemma leads to the following theorem.

Theorem

$$\sum_{n=1}^{N} a_n m_n q(j - m_n) = jq(j), \quad 1 \leq j \leq s \tag{A.65}$$

where $q(j) = 0$ for $j < 0$.

Proof. Multiplying both sides of (A.55) by m_n and summing over n yields

$$\sum_{n=1}^{N} a_n m_n q(j - m_n) = q(j) \sum_{n=1}^{N} m_n E[k_n \mid \boldsymbol{m} \cdot \boldsymbol{k} = j]$$

$$= q(j) E\left[\sum_{n=1}^{N} m_n k_n \mid \boldsymbol{m} \cdot \boldsymbol{k} = j\right]$$

$$= jq(j). \tag{A.66}$$

q. e. d.

Finally we obtain the following recursive relation for $\{G(j) \, 0 \leq j \leq s\}$.

Corollary

$$jG(j) = \sum_{l=0}^{j-1} G(l) + \sum_{n=1}^{N} a_n m_n G(j - m_n), \quad 1 \leq j \leq s. \tag{A.67}$$

Proof. From (A.65), we have for $1 \leq l \leq s$

$$G(s) \sum_{n=1}^{N} a_n m_n q(l - m_n) = l \cdot G(s) q(l) = l[G(l) - G(l - 1)]$$

$$= lG(l) - (l - 1)G(l - 1) - G(l - 1). \tag{A.68}$$

Summing the both sides over $l = 1, 2, \ldots, j$ we obtain

$$\sum_{n=1}^{N} a_n m_n G(j - m_n) = jG(j) - \sum_{l=0}^{j-1} G(l). \tag{A.69}$$

q. e. d.

The recursive formula in Equation (A.65) is usually called *Kaufman–Roberts algorithm* (Kaufman 1981; Roberts 1981), although it was first derived by Fortet and Grandjean (1964). Equation (A.67) is shown by Takagi *et al.* (2006). It is used in the methodology of required capacity calculation for the circuit-switched service categories in IMT-Advanced systems in Section 4.3.4.

A.4 M/G/1 Nonpreemptive Priority Queue

We consider a single server queue with an infinite waiting room. There are N classes of calls indexed $n = 1, 2, \ldots, N$. Calls of class n arrive in a Poisson process at rate λ_n. The service times of calls of class n have general distribution with average b_n and second moment $b_n^{(2)}$. We assume that the classes of calls are *priority classes* with the descending order of indices. That is, class n has priority over class i if and only if $n < i$. Class 1 is the highest priority class and class N the lowest. When the server becomes available, it selects for service one call of which priority is the highest among those calls present in the waiting room. We assume the FCFS discipline for the order of service within the same class of calls. The service is *nonpreemptive* in the sense that once the service to a call is started it is not disrupted until the completion (even if any calls of higher priority arrive during the service).

We calculate the average waiting time for calls of class n, denoted by W_n for $1 \leq n \leq N$ (Kleinrock 1976, pp. 119–121). To do so, we focus on a newly arriving call of class n which is referred to as a 'tagged' call. The waiting time of this tagged call consists of the following components:

- delay due to the call in service when our tagged call arrives;

- delay due to those calls found in the waiting room that receive service before our tagged call;

- delay due to those calls which arrive at the system after our tagged call but receive service before the tagged call.

Consequently, the average waiting time for our tagged call can be written as

$$W_n = W_0 + \sum_{i=1}^{N} b_i (L_{i,n} + M_{i,n}), \quad 1 \leq n \leq N, \tag{A.70}$$

where

$$W_0 := \text{average delay to our tagged call due to the call}$$
$$\text{found in service when our tagged call arrives} \tag{A.71}$$

$$L_{i,n} := \text{average number of calls of class } i \text{ found in the waiting room}$$
$$\text{by our tagged call of class } n \text{ and which receive service before}$$
$$\text{the tagged call} \tag{A.72}$$

$$M_{i,n} := \text{average number of calls of class } i \text{ that arrive at the system}$$
$$\text{while our tagged call of class } n \text{ is in the waiting room and}$$
$$\text{which receive service before the tagged call.} \tag{A.73}$$

We first evaluate W_0. The delay to our tagged call due to a call of class i found in service by the tagged call is the remaining service time of the call of class i when the tagged call arrives. The average remaining service time of a call of class i is derived as follows (Kleinrock 1975, pp. 169–173).

Let $f_i(x)$ be the probability density function (pdf) for the service time of a call of class i. It is important to note that the pdf $\hat{f}_i(x)$ for the service time $\hat{X}_i$ of a call of class i during which a call arrives is generally different from $f_i(x)$. This is because long service times occupy larger segments of the time axis than short service times do, and therefore it is more likely that an arbitrary call arrives during a long service time. The probability that a call arrives during the service time of a call of class i with duration x is proportional to the duration x as well as to the probability density $f_i(x)$. Thus

$$\hat{f}_i(x) = cxf_i(x), \tag{A.74}$$

where c is a constant. We can obtain $c = 1/b_i$ from the normalization condition as

$$1 = \int_0^\infty \hat{f}_i(x)\,dx = c\int_0^\infty xf_i(x)\,dx = cb_i. \tag{A.75}$$

Hence we obtain

$$\hat{f}_i(x) = \frac{xf_i(x)}{b_i}. \tag{A.76}$$

Let us now find the pdf $g_i(y)$ for the remaining service time Y_i of a call of class i. Given that our tagged call arrives during the service time of a call of class i with duration x, the arrival point is uniformly distributed over an interval $[0, x]$. Then we have

$$P\{Y_i \le y\} = \frac{y}{x}, \quad 0 \le y \le x. \tag{A.77}$$

Hence the joint density of $\hat{X}_i$ and Y_i is given by

$$P\{y < Y_i \le y + dy, x < \hat{X}_i \le x + dx\} = \frac{dy}{x} \cdot \frac{xf_i(x)\,dx}{b_i} = \frac{f_i(x)\,dy\,dx}{b_i}, \quad 0 \le y \le x. \tag{A.78}$$

Integrating over x, we obtain the pdf for Y_i as

$$g_i(y) = \int_{x=y}^{x=\infty} \frac{f_i(x)\,dx}{b_i} = \frac{1}{b_i}\int_y^\infty f_i(x)\,dx = \frac{1 - F_i(y)}{b_i} \tag{A.79}$$

where $F_i(y) := \int_0^y f_i(x)\,dx$ is the distribution function of X_i. We then obtain

$$E[Y_i] = \int_0^\infty yg_i(y)\,dy = \frac{1}{b_i}\int_0^\infty y[1 - F_i(y)]\,dy = \frac{b_i^{(2)}}{2b_i} \tag{A.80}$$

where the last result can be calculated by the method of partial integration.

With the above preparation, we can proceed to obtain W_0. To do so, we note that

$$\rho_i := \lambda_i b_i \tag{A.81}$$

is the fraction of time that the server is occupied by calls of class i. From the PASTA property, this is also the probability that our tagged call finds a call of class i in service. The average time until the completion of such service is given by $b_i^{(2)}/(2b_i)$. Hence we obtain

$$W_0 = \left(1 - \sum_{i=1}^N \rho_i\right) \cdot 0 + \sum_{i=1}^N \rho_i \cdot \frac{b_i^{(2)}}{2b_i} = \frac{1}{2}\sum_{i=1}^N \lambda_i b_i^{(2)}, \tag{A.82}$$

where the factor $1 - \sum_{i=1}^{N} \rho_i$ in the middle expression is the probability that there are no calls in service when our tagged call arrives.

We next evaluate $L_{i,n}$. It is clear that those calls of classes $n + 1, n + 2, \ldots, N$ (lower priority) found in the waiting room by our tagged call do not delay the tagged call. Hence

$$L_{i,n} = 0, \quad n + 1 \leq i \leq N. \tag{A.83}$$

On the other hand, those calls of classes $1, 2, \ldots, n$ (higher or equal priority) which are present in the waiting room upon the arrival of our tagged call must certainly be served before the tagged call. It follows from Little's law that the average number of such calls of class i present in the waiting room at an arbitrary time is given by $\lambda_i W_i$. From PASTA, this is also true at the arrival time of the tagged call. Thus we have

$$L_{i,n} = \lambda_i W_i, \quad 1 \leq i \leq n. \tag{A.84}$$

We finally evaluate $M_{i,n}$. Not only those calls of classes $n + 1, n + 2, \ldots, N$ (lower priority) but also the calls of class n (the same priority) that arrive while our tagged call is in the waiting room do not delay the tagged call. The latter is due to the assumption that calls within the same class are served according to the FCFS discipline. Hence

$$M_{i,n} = 0, \quad n \leq i \leq N. \tag{A.85}$$

Those calls of classes $1, 2, \ldots, n - 1$ (higher priority) that arrive while our tagged call is in the waiting room will also be served before the tagged call. The average time that the tagged call spends in the waiting room is W_n. Since the arrival rate of calls of class i is λ_i and since the arrival process of each class is independent of the number of calls in the waiting room, there will be on average $\lambda_i W_n$ arrivals of calls of class i while our tagged call is in the waiting room. Thus we have

$$M_{i,n} = \lambda_i W_n, \quad 1 \leq i \leq n - 1. \tag{A.86}$$

Note that this is *not* from Little's law.

Substituting Equations (A.82) to (A.86) into Equation (A.70), we obtain

$$
\begin{aligned}
W_n &= W_0 + \sum_{i=1}^{n} b_i \lambda_i W_i + \sum_{i=1}^{n-1} b_i \lambda_i W_n \\
&= W_0 + \sum_{i=1}^{n} \rho_i W_i + W_n \sum_{i=1}^{n-1} \rho_i \\
&= W_0 + \sum_{i=1}^{n-1} \rho_i W_i + W_n \sum_{i=1}^{n} \rho_i, \quad 1 \leq n \leq N,
\end{aligned}
\tag{A.87}
$$

which can be written as

$$W_n = \frac{W_0 + \sum_{i=1}^{n-1} \rho_i W_i}{1 - \sum_{i=1}^{n} \rho_i}, \quad 1 \leq n \leq N, \tag{A.88}$$

where the null sum $\sum_{i=1}^{0}$ is zero. This triangular set of simultaneous equations can be solved recursively by starting with W_1, then calculating W_2, and so on to obtain

$$
\begin{aligned}
W_n &= \frac{W_0}{(1 - \sum_{i=1}^{n-1} \rho_i)(1 - \sum_{i=1}^{n} \rho_i)} \\
&= \frac{\sum_{i=1}^{N} \lambda_i b_i^{(2)}}{2(1 - \sum_{i=1}^{n-1} \lambda_i b_i)(1 - \sum_{i=1}^{n} \lambda_i b_i)}, \quad 1 \le n \le N.
\end{aligned}
\tag{A.89}
$$

This formula was first derived by Cobham (1954). It is used in the methodology of required capacity calculation for the packet-switched service categories in IMT-Advanced systems in Section 4.3.5.

We may confirm that the inequality

$$
W_{n-1} < W_n, \quad 1 \le n \le N,
\tag{A.90}
$$

which demonstrates the effect of prioritized handling. It is also interesting to note the relation

$$
\sum_{n=1}^{N} \rho_n W_n = \frac{\rho W_0}{1 - \rho}
\tag{A.91}
$$

where $\rho := \sum_{n=1}^{N} \rho_n$. This can be easily seen from

$$
\frac{\rho_n}{(1 - \sum_{i=1}^{n-1} \rho_i)(1 - \sum_{i=1}^{n} \rho_i)} = \frac{1}{1 - \sum_{i=1}^{n} \rho_i} - \frac{1}{1 - \sum_{i=1}^{n-1} \rho_i}, \quad 1 \le n \le N.
\tag{A.92}
$$

The relation in (A.91) holds for any M/G/1 queue with multiple classes that has nonpreemptive work-conserving service discipline (Kleinrock 1976, pp. 113–117). This relationship is called *Kleinrock's conservation law*.

Appendix B

Example Market Study Parameter Values

The market study parameters shown in the following are:

- user density $U_{m,n}$ (users/km^2)

- session arrival rate per user $Q_{m,n}$ (sessions/h/user)

- average session duration $\mu_{m,n}$ (s/session)

- mean service bit rate $r_{m,n}$ (kbit/s)

- mobility ratios (%, dimensionless),

where m is the index for service environment (SE) and n is the index for service category (SC). We note that the unit 'sessions/h/user' for $Q_{m,n}$ is different from 'sessions/s/user' and that the unit 'kbit/s' for $r_{m,n}$ is different from 'bit/s' that are used in other places of the book. See Sections 4.2.1, 6.2 and 8.2.

We show the market study parameter values of the higher user density scenario in the forecast year 2020 for the following transmission modes and link directions:

- unicast downlink (Tables B.1–B.3)

- unicast uplink (Tables B.4–B.6)

- multicast downlink (Table B.7).

Originally, Tables B.1–B.3 are Tables 29b, Tables B.4–B.6 are Tables 29c, and Table B.7 is Table 29d, respectively, in Report ITU-R M.2078. (The order of columns for $\mu_{m,n}$ and $r_{m,n}$ are changed from the tables in Report ITU-R M.2078 in accordance with the order of description in Chapter 6.) However, the authors of this book found that some values in the tables of Report ITU-R M.2078 are incorrect. Tables B.1–B.7 below show the correct values obtained from the calculation tool.

Spectrum Requirement Planning in Wireless Communications   Edited by Hideaki Takagi and Bernhard H. Walke
© 2008 John Wiley & Sons, Ltd

Table B.1　Market study parameter values in the year 2020 for unicast downlink (higher user density case) (part 1 of 3).

SC n	SE m	$U_{m,n}$	$Q_{m,n}$	$\mu_{m,n}$	$r_{m,n}$	Mobility ratio			
						stationary	low	high	super-high
1	1	0.0	0.0	0.0	0.0	0.0	0.0	0.0	0.0
1	2	0.0	0.0	0.0	0.0	0.0	0.0	0.0	0.0
1	3	0.0	0.0	0.0	0.0	0.0	0.0	0.0	0.0
1	4	0.0	0.0	0.0	0.0	0.0	0.0	0.0	0.0
1	5	0.0	0.0	0.0	0.0	0.0	0.0	0.0	0.0
1	6	0.0	0.0	0.0	0.0	0.0	0.0	0.0	0.0
2	1	4730.0	0.144	360.0	11 240.0	55.0	25.0	20.0	0.0
2	2	17 062.5	0.707	513.3	11 240.0	77.5	17.5	5.0	0.0
2	3	10 216.8	0.144	360.0	11 240.0	10.0	70.0	20.0	0.0
2	4	1419.0	0.144	360.0	11 240.0	45.0	30.0	25.0	0.0
2	5	3789.3	0.707	508.4	11 240.0	55.0	10.0	30.0	5.0
2	6	283.8	0.144	360.0	11 240.0	5.0	10.0	75.0	15.0
3	1	18 096.0	0.384	231.3	506.0	69.0	21.0	10.0	0.0
3	2	35 528.0	0.630	214.8	379.8	69.0	26.0	5.0	0.0
3	3	26 291.5	0.301	146.9	470.9	47.0	43.0	10.0	0.0
3	4	2338.5	0.888	229.2	290.2	65.5	22.0	12.5	0.0
3	5	5266.3	1.04	172.2	275.2	49.0	16.0	30.0	5.0
3	6	387.8	0.299	161.1	282.3	46.0	11.5	35.0	7.5
4	1	13 089.0	0.995	810.7	88.0	73.8	11.9	9.5	4.8
4	2	13 128.0	0.995	810.7	88.0	73.1	16.5	5.7	4.7
4	3	17 421.0	1.02	833.8	123.0	49.3	32.3	13.8	4.6
4	4	14.3	1.21	810.7	88.0	69.0	14.3	11.9	4.8
4	5	33.8	1.13	819.8	105.0	54.8	9.5	30.7	5.0
4	6	14.5	1.18	814.2	95.3	50.7	8.2	33.8	7.2
5	1	37 575.8	0.925	229.0	16.0	62.6	22.4	9.3	5.6
5	2	68 203.0	1.33	227.2	16.0	62.0	25.5	6.9	5.6
5	3	45 589.3	1.00	252.1	15.3	37.7	42.5	14.2	5.7
5	4	4373.8	1.70	204.3	15.3	59.0	23.6	11.8	5.7
5	5	8709.5	2.34	255.1	11.8	34.1	23.4	36.6	5.9
5	6	842.8	1.68	203.6	15.3	36.1	18.3	36.1	9.6
6	1	1743.0	0.025	150.0	321 000.0	55.0	25.0	20.0	0.0
6	2	1743.0	0.025	150.0	321 000.0	55.0	35.0	10.0	0.0
6	3	2324.0	0.030	150.0	321 000.0	10.0	70.0	20.0	0.0
6	4	0.0	0.0	0.0	0.0	0.0	0.0	0.0	0.0
6	5	0.0	0.0	0.0	0.0	0.0	0.0	0.0	0.0
6	6	0.0	0.0	0.0	0.0	0.0	0.0	0.0	0.0
7	1	5080.5	0.017	892.8	3075.0	55.0	25.0	20.0	0.0
7	2	13 683.0	0.639	1136.3	8075.5	70.0	21.7	5.8	2.4
7	3	2971.8	0.167	480.0	10 963.1	33.3	42.3	17.9	6.5
7	4	917.0	0.023	1080.0	3000.0	45.0	30.0	25.0	0.0
7	5	44.3	1.032	533.6	9992.8	48.5	14.1	30.1	7.3
7	6	187.0	0.023	988.2	3409.0	6.5	10.0	68.5	15.0

Table B.2   Market study parameter values in the year 2020 for unicast downlink (higher user density case) (part 2 of 3).

SC n	SE m	$U_{m,n}$	$Q_{m,n}$	$\mu_{m,n}$	$r_{m,n}$	Mobility ratio			
						stationary	low	high	super-high
8	1	15782.5	0.174	123.0	700.8	72.5	17.5	10.0	0.0
8	2	15861.0	0.932	486.2	868.8	63.9	19.8	8.8	7.5
8	3	21320.3	0.793	486.2	868.8	46.1	36.9	9.2	7.8
8	4	21.0	0.411	158.0	384.0	90.0	10.0	0.0	0.0
8	5	68.5	1.03	493.1	868.8	47.6	16.7	28.6	7.1
8	6	9.0	0.739	493.1	868.8	39.7	15.4	37.4	7.5
9	1	309.0	0.823	20.0	144.0	90.0	10.0	0.0	0.0
9	2	412.0	1.23	79.0	144.0	90.0	10.0	0.0	0.0
9	3	52.0	0.411	79.0	144.0	80.0	10.0	10.0	0.0
9	4	52.0	0.823	20.0	144.0	90.0	10.0	0.0	0.0
9	5	103.0	1.23	79.0	144.0	70.0	10.0	20.0	0.0
9	6	10.0	0.411	20.0	144.0	80.0	10.0	10.0	0.0
10	1	206.0	0.823	4.0	16.0	80.0	20.0	0.0	0.0
10	2	309.0	1.23	4.0	16.0	80.0	20.0	0.0	0.0
10	3	52.0	0.411	4.0	16.0	70.0	20.0	10.0	0.0
10	4	52.0	0.823	4.0	16.0	80.0	20.0	0.0	0.0
10	5	52.0	1.23	4.0	16.0	60.0	20.0	20.0	0.0
10	6	10.0	0.411	4.0	16.0	65.0	20.0	10.0	5.0
11	1	27.8	0.030	18.0	321000.0	55.0	25.0	20.0	0.0
11	2	43.5	0.317	81.9	321000.0	100.0	0.0	0.0	0.0
11	3	18793.8	0.299	6.0	90141.2	10.0	70.0	20.0	0.0
11	4	0.0	0.0	0.0	0.0	0.0	0.0	0.0	0.0
11	5	6950.8	0.876	53.9	212984.6	100.0	0.0	0.0	0.0
11	6	521.0	0.300	6.0	89978.0	5.0	10.0	70.0	15.0
12	1	35244.0	2.84	33.1	11059.6	67.5	22.5	10.0	0.0
12	2	120974.5	2.57	75.5	10725.9	67.5	27.5	5.0	0.0
12	3	82235.8	2.28	75.2	11344.8	45.0	45.0	10.0	0.0
12	4	10549.5	2.85	21.2	9835.8	62.5	25.0	12.5	0.0
12	5	30422.0	2.31	21.2	9858.5	42.5	20.0	32.5	5.0
12	6	2373.3	2.56	21.2	9953.7	40.0	15.0	37.5	7.5
13	1	14325.3	0.243	149.3	1360.5	69.3	20.7	6.7	3.3
13	2	14293.5	0.503	148.6	1360.8	69.1	22.1	5.4	3.4
13	3	17566.0	0.329	156.3	1359.6	40.3	38.1	16.5	5.1
13	4	75.5	0.970	176.6	1358.1	57.5	23.4	11.7	7.5
13	5	110.8	1.76	175.2	1358.1	38.6	19.8	33.8	7.7
13	6	24.5	1.14	177.9	1249.6	35.5	15.3	39.4	9.9
14	1	15514.5	0.240	30.6	120.2	67.2	22.4	10.0	0.5
14	2	15514.5	0.254	30.3	121.6	67.2	27.4	5.0	0.5
14	3	26031.8	0.312	314.1	177.0	26.5	61.0	11.5	1.0
14	4	67.3	0.528	66.9	48.8	61.3	22.1	12.3	4.4
14	5	2977.3	0.297	981.0	177.8	35.2	59.8	5.0	0.0
14	6	233.3	0.323	852.6	177.6	33.0	58.5	5.5	3.0

Table B.3   Market study parameter values in the year 2020 for unicast downlink (higher user density case) (part 3 of 3).

SC	SE					Mobility ratio			
n	m	$U_{m,n}$	$Q_{m,n}$	$\mu_{m,n}$	$r_{m,n}$	stationary	low	high	super-high
15	1	13 139.8	2.39	30.0	12.5	65.9	21.5	9.8	2.9
15	2	41 089.8	3.44	33.2	9.7	66.2	27.0	5.4	1.5
15	3	25 608.0	1.87	35.9	28.3	33.0	39.8	23.8	3.4
15	4	3382.0	2.67	27.2	9.0	62.2	24.9	12.4	0.5
15	5	9313.8	3.43	29.3	58.3	43.5	12.5	40.0	4.0
15	6	702.8	2.65	30.7	77.5	34.5	14.5	40.5	10.5
16	1	0.0	0.0	0.0	0.0	0.0	0.0	0.0	0.0
16	2	52.0	0.823	222.0	20 000.0	80.0	20.0	0.0	0.0
16	3	0.0	0.0	0.0	0.0	0.0	0.0	0.0	0.0
16	4	0.0	0.0	0.0	0.0	0.0	0.0	0.0	0.0
16	5	10.0	0.823	222.0	20 000.0	80.0	20.0	0.0	0.0
16	6	0.0	0.0	0.0	0.0	0.0	0.0	0.0	0.0
17	1	4493.3	0.713	32.7	10 056.5	56.0	24.0	20.0	0.0
17	2	16 242.8	1.01	179.5	9931.2	67.5	27.5	5.0	0.0
17	3	2969.5	0.824	90.6	10 911.4	12.0	68.0	19.5	0.5
17	4	1309.8	0.728	32.1	9481.7	45.0	30.0	25.0	0.0
17	5	1044.0	1.43	224.9	9733.7	45.0	20.0	30.0	5.0
17	6	262.5	0.727	32.7	9816.7	6.0	10.0	69.0	15.0
18	1	618.0	0.411	82.0	574.0	80.0	20.0	0.0	0.0
18	2	1339.0	4.11	153.0	595.0	80.0	20.0	0.0	0.0
18	3	21.0	0.411	23.0	990.0	60.0	20.0	20.0	0.0
18	4	103.0	0.411	72.0	1030.0	80.0	20.0	0.0	0.0
18	5	72.0	4.11	148.0	653.0	60.0	20.0	20.0	0.0
18	6	10.0	0.411	72.0	1030.0	60.0	20.0	15.0	0.0
19	1	515.0	0.411	21.0	144.0	80.0	20.0	0.0	0.0
19	2	1030.0	1.23	21.0	144.0	80.0	20.0	0.0	0.0
19	3	52.0	0.411	21.0	144.0	60.0	20.0	15.0	5.0
19	4	52.0	0.411	21.0	144.0	80.0	20.0	0.0	0.0
19	5	52.0	0.411	21.0	144.0	60.0	20.0	20.0	0.0
19	6	10.0	0.411	21.0	144.0	60.0	20.0	15.0	5.0
20	1	1030.0	0.411	25.0	16.0	80.0	20.0	0.0	0.0
20	2	1030.0	1.23	41.0	16.0	80.0	20.0	0.0	0.0
20	3	103.0	0.411	25.0	16.0	60.0	20.0	20.0	0.0
20	4	103.0	0.411	25.0	16.0	80.0	20.0	0.0	0.0
20	5	52.0	1.23	41.0	16.0	60.0	20.0	20.0	0.0
20	6	10.0	0.411	25.0	16.0	60.0	20.0	15.0	5.0

Table B.4   Market study parameter values in the year 2020 for unicast uplink (higher user density case) (part 1 of 3).

SC n	SE m	$U_{m,n}$	$Q_{m,n}$	$\mu_{m,n}$	$r_{m,n}$	Mobility ratio stationary	low	high	super-high
1	1	0.0	0.0	0.0	0.0	0.0	0.0	0.0	0.0
1	2	0.0	0.0	0.0	0.0	0.0	0.0	0.0	0.0
1	3	0.0	0.0	0.0	0.0	0.0	0.0	0.0	0.0
1	4	0.0	0.0	0.0	0.0	0.0	0.0	0.0	0.0
1	5	0.0	0.0	0.0	0.0	0.0	0.0	0.0	0.0
1	6	0.0	0.0	0.0	0.0	0.0	0.0	0.0	0.0
2	1	52.0	0.411	7406.0	20000.0	100.0	0.0	0.0	0.0
2	2	57.0	1.16	265.0	20000.0	100.0	0.0	0.0	0.0
2	3	0.0	0.0	0.0	0.0	0.0	0.0	0.0	0.0
2	4	21.0	0.411	7406.0	20000.0	100.0	0.0	0.0	0.0
2	5	10.0	1.23	35.0	20000.0	100.0	0.0	0.0	0.0
2	6	1.0	0.411	7406.0	20000.0	100.0	0.0	0.0	0.0
3	1	18389.3	0.384	252.6	496.4	69.0	21.0	10.0	0.0
3	2	35875.3	0.629	221.1	323.4	69.5	25.5	5.0	0.0
3	3	26338.0	0.301	180.5	397.1	46.5	43.5	10.0	0.0
3	4	2415.8	0.880	245.1	275.2	66.0	21.5	12.5	0.0
3	5	5309.0	1.01	196.2	275.2	49.0	16.0	30.0	5.0
3	6	389.3	0.294	175.5	275.2	46.0	11.5	35.0	7.5
4	1	13089.0	0.995	3032.5	88.0	73.8	11.9	9.5	4.8
4	2	13128.0	0.995	3032.5	88.0	73.1	16.5	5.7	4.7
4	3	17421.0	1.02	3055.6	1996.8	49.3	32.3	13.8	4.6
4	4	14.3	1.21	3032.5	88.0	69.0	14.3	11.9	4.8
4	5	33.8	1.13	3041.6	985.5	54.8	9.5	30.7	5.0
4	6	14.5	1.18	3036.0	492.9	50.7	8.2	33.8	7.2
5	1	37575.8	0.925	229.0	16.0	62.6	22.4	9.3	5.6
5	2	68203.0	1.33	227.2	16.0	62.0	25.5	6.9	5.6
5	3	45589.3	1.00	252.1	15.3	37.7	42.5	14.2	5.7
5	4	4373.8	1.70	204.3	15.3	59.0	23.6	11.8	5.7
5	5	8709.5	2.34	255.1	11.8	34.1	23.4	36.6	5.9
5	6	842.8	1.68	203.6	15.3	36.1	18.3	36.1	9.6
6	1	1743.0	0.025	150.0	302293.9	55.0	25.0	20.0	0.0
6	2	1743.0	0.025	150.0	302293.9	55.0	35.0	10.0	0.0
6	3	2324.0	0.030	150.0	302293.9	10.0	70.0	20.0	0.0
6	4	0.0	0.0	0.0	0.0	0.0	0.0	0.0	0.0
6	5	0.0	0.0	0.0	0.0	0.0	0.0	0.0	0.0
6	6	0.0	0.0	0.0	0.0	0.0	0.0	0.0	0.0
7	1	11366.8	0.179	350.4	3201.5	55.0	25.0	20.0	0.0
7	2	35845.3	1.10	407.0	8233.0	72.5	22.5	5.0	0.0
7	3	23151.5	0.215	358.8	3688.5	11.3	68.5	19.7	0.5
7	4	2802.8	0.217	352.8	3194.4	45.0	30.0	25.0	0.0
7	5	7518.3	1.12	404.6	8233.7	50.0	15.0	30.0	5.0
7	6	564.3	0.216	353.1	3240.7	5.0	10.0	70.0	15.0

Table B.5   Market study parameter values in the year 2020 for unicast uplink (higher user density case) (part 2 of 3).

| SC | SE | | | | | Mobility ratio | | | |
n	m	$U_{m,n}$	$Q_{m,n}$	$\mu_{m,n}$	$r_{m,n}$	stationary	low	high	super-high
8	1	15 782.5	0.174	59.7	653.2	72.5	17.5	10.0	0.0
8	2	15 861.0	0.932	402.9	868.8	63.9	19.8	8.8	7.5
8	3	21 320.3	0.793	402.9	868.8	46.1	36.9	9.2	7.8
8	4	21.0	0.411	39.0	384.0	90.0	10.0	0.0	0.0
8	5	68.5	1.03	409.8	868.8	47.6	16.7	28.6	7.1
8	6	9.0	0.739	409.8	868.8	39.7	15.4	37.4	7.5
9	1	309.0	0.823	30.0	144.0	90.0	10.0	0.0	0.0
9	2	412.0	1.23	118.0	144.0	90.0	10.0	0.0	0.0
9	3	52.0	0.411	118.0	144.0	80.0	10.0	10.0	0.0
9	4	52.0	0.823	30.0	144.0	90.0	10.0	0.0	0.0
9	5	103.0	1.23	118.0	144.0	70.0	10.0	20.0	0.0
9	6	10.0	0.411	30.0	144.0	80.0	10.0	10.0	0.0
10	1	2934.5	0.876	1086.3	11.8	67.5	22.5	10.0	0.0
10	2	10 239.8	1.07	1086.3	11.8	67.5	27.5	5.0	0.0
10	3	6043.8	0.588	1086.3	11.8	40.0	45.0	15.0	0.0
10	4	873.0	0.876	1086.3	11.8	62.5	25.0	12.5	0.0
10	5	2263.0	1.07	1086.3	11.8	35.0	20.0	40.0	5.0
10	6	174.3	0.588	1086.3	11.8	35.0	15.0	40.0	10.0
11	1	27.8	0.030	18.0	304 587.8	55.0	25.0	20.0	0.0
11	2	43.5	0.317	28.5	304 587.8	100.0	0.0	0.0	0.0
11	3	37.0	0.036	18.0	304 587.8	10.0	70.0	20.0	0.0
11	4	0.0	0.0	0.0	0.0	0.0	0.0	0.0	0.0
11	5	5.0	0.823	15.0	500 000.0	100.0	0.0	0.0	0.0
11	6	0.0	0.0	0.0	0.0	0.0	0.0	0.0	0.0
12	1	22 068.5	0.342	20.7	10 530.7	67.7	22.4	9.5	0.5
12	2	78 431.0	0.341	19.8	8732.8	67.5	27.5	5.0	0.0
12	3	47 267.3	0.345	21.0	11 638.4	49.0	41.5	9.0	0.5
12	4	6539.3	0.341	19.8	8680.9	62.5	25.0	12.5	0.0
12	5	17 376.0	0.341	19.8	8817.1	43.0	20.0	32.0	5.0
12	6	1305.3	0.342	20.1	9469.5	41.3	15.4	35.8	7.5
13	1	15 821.3	1.25	34.9	1074.5	69.3	20.7	6.7	3.3
13	2	14 293.5	0.503	89.5	1030.9	69.1	22.1	5.4	3.4
13	3	17 566.0	0.329	97.2	1181.6	39.8	38.1	17.0	5.1
13	4	75.5	8.87	32.1	1187.0	57.7	23.5	11.7	7.0
13	5	114.5	8.47	110.9	1216.7	38.6	19.8	33.8	7.7
13	6	24.5	8.47	35.2	1121.1	35.1	15.3	39.6	9.9
14	1	15 514.5	0.240	27.1	59.3	67.2	22.4	10.0	0.5
14	2	15 514.5	0.254	26.8	59.3	67.2	27.4	5.0	0.5
14	3	26 031.8	0.312	310.6	99.1	26.5	61.0	11.5	1.0
14	4	67.3	0.528	63.4	48.8	61.3	22.1	12.3	4.4
14	5	2977.3	0.297	977.5	92.7	35.2	59.8	5.0	0.0
14	6	233.3	0.323	849.1	94.8	33.0	58.5	5.5	3.0

Table B.6   Market study parameter values in the year 2020 for unicast uplink (higher user density case) (part 3 of 3).

SC n	SE m	$U_{m,n}$	$Q_{m,n}$	$\mu_{m,n}$	$r_{m,n}$	Mobility ratio			
						stationary	low	high	super-high
15	1	32 639.8	13.4	36.1	9.7	67.2	22.4	10.0	0.5
15	2	111 289.8	14.8	45.1	9.7	67.5	27.5	5.0	0.0
15	3	66 073.0	12.9	31.1	11.1	35.0	45.0	18.5	1.5
15	4	9232.0	13.7	36.1	9.0	62.5	25.0	12.5	0.0
15	5	24 300.8	15.0	36.1	9.7	45.0	15.0	35.0	5.0
15	6	1826.8	13.8	36.1	9.7	35.0	15.0	40.0	10.0
16	1	0.0	0.0	0.0	0.0	0.0	0.0	0.0	0.0
16	2	52.0	0.823	25.0	20 000.0	80.0	20.0	0.0	0.0
16	3	0.0	0.0	0.0	0.0	0.0	0.0	0.0	0.0
16	4	0.0	0.0	0.0	0.0	0.0	0.0	0.0	0.0
16	5	10.0	0.823	25.0	20 000.0	80.0	20.0	0.0	0.0
16	6	0.0	0.0	0.0	0.0	0.0	0.0	0.0	0.0
17	1	316.3	0.545	93.6	14 803.0	59.0	22.0	17.0	2.0
17	2	737.0	1.34	132.7	13 099.8	67.2	27.4	5.0	0.5
17	3	548.3	0.645	116.4	17 120.2	20.4	59.7	18.4	1.5
17	4	56.5	0.770	90.9	11 418.8	46.5	29.0	24.0	0.5
17	5	147.3	1.39	126.1	11 228.5	45.0	20.0	30.0	5.0
17	6	12.0	0.749	99.9	13 532.0	11.0	10.5	64.0	14.5
18	1	618.0	0.411	35.0	574.0	80.0	20.0	0.0	0.0
18	2	1339.0	4.11	66.0	595.0	80.0	20.0	0.0	0.0
18	3	21.0	0.411	10.0	990.0	60.0	20.0	20.0	0.0
18	4	103.0	0.411	31.0	1030.0	80.0	20.0	0.0	0.0
18	5	72.0	4.11	63.0	653.0	60.0	20.0	20.0	0.0
18	6	10.0	0.411	31.0	1030.0	60.0	20.0	15.0	5.0
19	1	515.0	0.411	21.0	144.0	80.0	20.0	0.0	0.0
19	2	1030.0	1.23	21.0	144.0	80.0	20.0	0.0	0.0
19	3	52.0	0.411	21.0	144.0	60.0	20.0	15.0	5.0
19	4	52.0	0.411	21.0	144.0	80.0	20.0	0.0	0.0
19	5	52.0	0.411	21.0	144.0	60.0	20.0	20.0	0.0
19	6	10.0	0.411	21.0	144.0	60.0	20.0	15.0	5.0
20	1	1030.0	0.411	25.0	16.0	80.0	20.0	0.0	0.0
20	2	1030.0	1.23	41.0	16.0	80.0	20.0	0.0	0.0
20	3	1211.3	0.588	8.2	13.2	35.0	45.0	20.0	0.0
20	4	103.0	0.411	25.0	16.0	80.0	20.0	0.0	0.0
20	5	459.0	1.07	13.0	13.2	35.0	20.0	50.0	5.0
20	6	39.0	0.588	8.2	13.2	32.5	15.0	42.5	10.0

Table B.7   Market study parameter values in the year 2020 for multicast downlink (higher user density case).

SC n	SE m	$U_{m,n}$	$Q_{m,n}$	$\mu_{m,n}$	$r_{m,n}$	Mobility ratio stationary	low	high	super-high
2	1	51.5	0.4	14 812.0	20 000.0	100.0	0.0	0.0	0.0
2	2	10.3	1.7	5554.5	20 000.0	100.0	0.0	0.0	0.0
2	3	0.0	0.0	0.0	0.0	0.0	0.0	0.0	0.0
2	4	20.6	0.4	14 812.0	20 000.0	100.0	0.0	0.0	0.0
2	5	3.1	1.7	2468.7	20 000.0	100.0	0.0	0.0	0.0
2	6	1.0	0.4	14 812.0	20 000.0	100.0	0.0	0.0	0.0
3	1	391.4	1.7	1130.4	1424.2	83.0	17.0	0.0	0.0
3	2	463.5	2.5	1587.0	922.7	84.0	16.0	0.0	0.0
3	3	61.8	1.7	617.2	1192.0	82.0	18.0	0.0	0.0
3	4	103.0	5.4	1026.4	731.7	86.0	14.0	0.0	0.0
3	5	56.7	5.8	1388.6	623.4	87.0	13.0	0.0	0.0
3	6	2.7	3.3	922.5	679.9	87.0	13.0	0.0	0.0

Appendix C

List of Acronyms and Symbols

Acronyms and symbols used in the methodology for IMT-2000 that appear in Chapter 3 are indicated as [IMT-2000]. Those of queueing theory that appear in Appendix A are indicated as [queueing theory]. Otherwise they are general or used in the methodology for IMT-Advanced. Physical units of the symbols are shown in parentheses when applicable.

C.1 Acronyms

AAC Advanced Audio Coding

AC-3 Audio Code number 3

AM Amplitude Modulation

AMPS Advanced Mobile Phone Service

APT Asia-Pacific Telecommunity

ARIB Association of Radio Industries and Businesses

ATSC Advanced Television Systems Committee

AVM Automatic Vehicle Monitoring

AWCSC Advanced Wireless Communications Study Committee

BCMCS BroadCast and MultiCast Services

BHCA Busy Hour Call Attempts [IMT-2000]

BS Base Station

BS Broadcast Service

BSC Base Station Controller

Spectrum Requirement Planning in Wireless Communications　Edited by Hideaki Takagi and Bernhard H. Walke
© 2008 John Wiley & Sons, Ltd

B3G Beyond the Third Generation

CATV Cable Television, Community Antenna Television

CBD Central Business District [IMT-2000]

CCSA China Communications Standard Association

CDM Code Division Multiplex

CDMA Code Division Multiple Access

CEPT Conférence Européenne des Administrations des Postes et des Télécommunications (European Conference of Postal and Telecommunications Administrations)

CITEL Comisión Interamericana de Telecomunicaciones (Inter-American Telecommunication Commission)

CJK China-Japan-Korea

CN Core Network

COFDM Coded Orthogonal Frequency Division Multiplexing

COST European Cooperation in the field of Scientific and Technical Research

CPM Conference Preparatory Meeting

DAB Digital Audio Broadcasting

DARPA Defense Advanced Research Projects Agency

dB Decibel

DECT Digital Enhanced Cordless Telecommunications

DF Distribution Function

DL DownLink

DO Data Only

DQPSK Differential Quadrature Phase Shift Keying

DSSS Direct Sequence Spread Spectrum

DVB-T Digital Video Broadcasting – Terrestrial

DySPAN Dynamic Spectrum Access Networks

EDGE Enhanced Data Rate for Global Evolution

EGPRS Enhanced General Packet Radio Service

EHF Extremely High Frequency

EO External Organization

ETSI European Telecommunication Standards Institute

EU European Union

EV-DO EVolution Data Only (Optimized)

FCFS First-Come First-Served [queueing theory]

FDD Frequency Division Duplex

FDM Frequency Division Multiplex

FDMA Frequency Division Multiple Access

FHSS Frequency Hopping Spread Spectrum

FM Frequency Modulation

FOMA Freedom Of Mobile Multimedia Access

FPLMTS Future Public Land Mobile Telecommunication System

FP5 Fifth Framework Programme

FP6 Sixth Framework Programme

FS Fixed Service

FSR Frequency Sharing Rule

FSS Fixed Satellite Service

FSU Flexible Spectrum Use

FuTURE Future Technologies for Universal Radio Environment

GERAN GSM EDGE Radio Access Network

GHz Giga Herz

GMSC Gateway Mobile Switching Center

GPRS General Packet Radio Service

GPS Global Positioning System

GSM Global System for Mobile communications

h Hour

HAPS High Altitude Platform Station

HF High Frequency

HIMM High Interactive MultiMedia [IMT-2000]

HLR Home Location Register

HMM High MultiMedia [IMT-2000]

HSDPA High Speed Downlink Packet Access

HSPA High Speed Packet Access

HSUPA High Speed Uplink Packet Access

HTSG High Throughput Study Group

Hz Herz

H_2 Hyperexponential distribution of the second order

IEEE Institute of Electrical and Electronics Engineers

IEEE SA IEEE Standards Association

IMS IP Multimedia Subsystem

IMT International Mobile Telecommunications

IMT-2000 International Mobile Telecommunications-2000

IMT-DS IMT-Direct Spread

IMT-FT IMT-Frequency Time

IMT-MC IMT-Multi Carrier

IMT-SC IMT-Single Carrier

IMT-TC IMT-Time Code

IP Internet Protocol

IR Infrared

ISDB-T Integrated Services Digital Broadcasting-Terrestrial

ISDN Integrated Services Digital Network

ISM Industrial, Scientific and Medical

IST Information Society Technologies

IS-95 Interim Standard-95

IT Information Technology

ITS Intelligent Transportation System

ITU International Telegraph Union [formerly]

ITU International Telecommunication Union

ITU-R International Telecommunication Union – Radiocommunication section

kbit/s Kilo bits per second

km Kilometer

LAN Local Area Network

LF Low Frequency

LORAN LOng RAnge Navigation system

LOS Line-Of-Sight

LST Laplace–Stieltjes Transform

LTE Long Term Evolution

MAC Medium Access Control layer

MAN Metropolitan Area Network

MB Mega Bytes

Mbit/s Megabits per second

MBMS Multimedia Broadcast/Multicast Service

M/D/1 Poisson arrival, deterministic service time queue with a single server

M/D*/1 Poisson arrival, multi-modal service time queue with a single server

MF Medium Frequency

M/G/1 Poisson arrival, generally distributed service time queue with a single server [queueing theory]

MHz Mega Hertz

M/{i D}/1 Poisson arrival, discrete service time queue with a single server

MIMO Multiple Input Multiple Output

MIND Mobile IP-based Network Developments project

mITF mobile IT Forum

MMM Medium MultiMedia [IMT-2000]

M/G/s/s Poisson arrival, generally distributed service time loss system with s servers [queueing theory]

M/M/s Poisson arrival, exponentially distributed service time delay system with s servers [queueing theory]

M/M/s/s Poisson arrival, exponentially distributed service time loss system with s servers [queueing theory]

MPEG-2 Moving Picture Experts Group-2

MS Microsoft

MS Mobile Service

MS Mobile Station

ms Millisecond

MSC Mobile Switching Center

MSS Mobile Satellite Service

MTU Maximum Transfer Unit

NGMC Next Generation Mobile Communications Forum

NMT Nordic Mobile Telephone

NWA Nomadic Wireless Access

OFDM Orthogonal Frequency Division Multiplexing

OFDMA Orthogonal Frequency Division Multiple Access

OMC Operation and Maintenance Center

PAR Project Authorization Request

PASTA Poisson Arrivals See Time Averages [queueing theory]

PC Personal Computer

PCS Personal Communications Service

PDA Personal Digital Assistant

PDC Personal Digital Cellular

pdf Probability density function [queueing theory]

PHS Personal Handy-phone System

PHY Physical layer

PMR Professional Mobile Radio

PSTN Public Switched Telephone Network

P2P Peer-to-Peer

QAM Quadrature Amplitude Modulation

q.e.d. quod erat demonstrandum (= which was to be proved) [queueing theory]

QoS Quality of Service

QPSK Quadrature Phase Shift Keying

RAM Random Access Memory

RAN Radio Access Network

RAT Radio Access Technique

RATG Radio Access Technique Group

RCR Research Center for Radio

RE Radio Environment

RLAN Radio Local Area Network

RR Radio Regulations

s Second

S Speech [IMT-2000]

SAP Service Access Point

SC Service Category

SCDMA Synchronous Code Division Multiple Access

SC-FDMA Single Carrier Frequency Division Multiple Access

SCV Squared Coefficient of Variation

SD Switched Data [IMT-2000]

SDO Standards Development Organization

SE Service Environment

SFN Single Frequency Network

SHF Super High Frequency

SG Study Group

SIG Special Interest Group

SM Simple Message [IMT-2000]

SMS Short Message Service

TACS Total Access Communication System

TAG Technical Advisory Group

TDD Time Division Duplex

TDM Time Division Multiplex

TDMA Time Division Multiple Access

TD-SCDMA Time Division Synchronous CDMA

TETRA TErrestrial Trunked RAdio

TG Task Group

THz Tera Hertz

TIA Telecommunications Industry Association

TTA Telecommunications Technology Association

TTC Telecommunication Technology Committee

TV Television

UHF Ultra High Frequency

UL UpLink

UMB Ultra Mobile Broadband

UMTS Universal Mobile Telecommunications System

UN United Nations

USB Universal Serial Bus

UTRA UMTS Terrestrial Radio Access

UWB Ultra Wide Band

UWC-136 Universal Wireless Communications-136

VHF Very High Frequency

VLF Very Low Frequency

VoIP Voice over Internet Protocol

VOR VHF Omnidirectional radio Range

WARC World Administrative Radiocommunication Conference

W-CDMA Wideband Code Division Multiple Access

WG Working Group

WiFi Wireless Fidelity

WiMAX Worldwide Interoperability for Microwave Access

WINNER Wireless World Initiative New Radio

WLAN Wireless Local Area Network

WP8F Working Party 8F

WRAN Wireless Regional Area Network

WRC World Radiocommunication Conference

xDSL x Digital Subscriber Line

XG neXt Generation communications program

1G First Generation

2G Second Generation

3G Third Generation

3GPP Third Generation Partnership Project

3GPP2 Third Generation Partnership Project 2

4G Fourth Generation

8-VSB 8-level Vestigial Side Band

C.2   Symbols

a Offered traffic (Erlang) [IMT-2000, queueing theory]

$A_{d,p}$ Cell area of radio environment p in teledensity d (km^2/cell)

$\texttt{Activity\_Factor}_{e,s,l}$ Percentage of time during which the link in direction l is actually used by service s in environment e (dimensionless) [IMT-2000]

b_n Mean service time of a packet of class n (s/packet)

$b_n^{(2)}$ Second moment of the service time of a packet of class n (s^2/packet^2)

$\texttt{Busy\_Hour\_Call\_Attempts}_{e,s}$ Average number of call attempts by a user of service s per hour during the busy hour in environment e (calls/h/user) [IMT-2000]

C Channel capacity (bit/s)

$C_{d,rat,p}$ Capacity requirement in teledensity d for RATG rat in radio environment p (bit/s/cell)

$C_{d,rat,p}^{cs}$ Capacity requirement for circuit-switched service categories in teledensity d for RATG rat in radio environment p (bit/s/cell)

$C_{d,rat,p}^{ps}$ Capacity requirement for packet-switched service categories in teledensity d for RATG rat in radio environment p (bit/s/cell)

Call_Duration$_{e,s}$ Average duration of a call/session of service s during the busy hour in environment e (s) [IMT-2000]

Cell_Area$_e$ Cell area of environment e (km^2/cell) [IMT-2000]

C_X^2 Squared coefficient of variation of the random variable X

d Index for teledensity

e Index for environment [IMT-2000]

$E[X]$ Expected value of the random variable X [queueing theory]

$E_B(s,a)$ Blocking probability in the Erlang-B formula for an M/M/s/s loss system with offered load a [IMT-2000, queueing theory]

$E_C(s,a)$ Wait probability in the Erlang-C formula for an M/M/s delay system with offered load a [IMT-2000, queueing theory]

F Total spectrum requirement (MHz) [IMT-2000]

$F_{d,rat}$ Spectrum requirement in teledensity d for RATG rat (Hz)

$F_{d,rat,p}$ Spectrum requirement in teledensity d for RATG rat in radio environment p (Hz)

$F_{d,rat,p}^{Multicast}$ Spectrum requirement for mobile multicast traffic in teledensity d for RATG rat in radio environment p (Hz)

$F_{e,s,l}$ Required spectrum for service s in environment e on link l (MHz) [IMT-2000]

F_l Required spectrum on link l (MHz) [IMT-2000]

F_{rat} Spectrum requirement for RATG rat (Hz)

F_{rat}^G Guard band between operators for RATG rat (Hz)

Group_Size$_{e,s}$ Number of cells per cell group (cells/group) [IMT-2000]

J_m Splitting factor in service environment m (dimensionless)

l Index for link direction [IMT-2000]

m Index for service environment

$\text{MinSpec}_{rat,p}$ Minimum bandwidth per operator and radio environment for RATG *rat* in radio environment *p* (Hz)

$\text{MR}_{\text{stat/ped};m,n}$ Stationary/pedestrian mobility ratio of service category *n* in service environment *m* (dimensionless)

$\text{MR}_{\text{low};m,n}$ Low mobility ratio of service category *n* in service environment *m* (dimensionless)

$\text{MR}_{\text{high};m,n}$ High mobility ratio of service category *n* in service environment *m* (dimensionless)

$\text{MR}_{m,s}^{\text{Market}}$ Market study mobility ratio of service *s* in service environment *m* (dimensionless)

$\text{MR}_{m,n}^{\text{Market}}$ Market study mobility ratio of service category *n* in service environment *m* (dimensionless)

$\text{MR}_{\text{stat};m,n}^{\text{Market}}$ Market study stationary mobility ratio of service category *n* in service environment *m* (dimensionless)

$\text{MR}_{\text{low};m,n}^{\text{Market}}$ Market study low mobility ratio of service category *n* in service environment *m* (dimensionless)

$\text{MR}_{\text{high};m,n}^{\text{Market}}$ Market study high mobility ratio of service category *n* in service environment *m* (dimensionless)

$\text{MR}_{\text{s-high};m,n}^{\text{Market}}$ Market study super-high mobility ratio of service category *n* in service environment *m* (dimensionless)

n Index for service category

N^{cs} Number of circuit-switched service categories

N^{ps} Number of packet-switched service categories

N_{o} Number of network operators

$\text{Net\_System\_Capability}_s$ Spectrum bandwidth needed to carry the data rate of 1 bit/s per cell (bit/s/Hz/cell) [IMT-2000]

$\text{Net\_User\_Bit\_Rate}_s$ Bit rate of bearer service *s* (kbit/s) [IMT-2000]

$\text{Offered\_Traffic\_per\_Cell}_{e,s,l}$ Offered traffic of service *s* per cell of environment *e* on link *l* (calls·seconds/hour/cell) [IMT-2000]

$\text{Offered\_Traffic\_per\_Group}_{e,s,l}$ Offered traffic of service *s* in a group of cells of environment *e* on link *l* (Erlang/group) [IMT-2000]

$\text{Offered\_Traffic\_per\_User}_{e,s,l}$ Offered traffic of service *s* by a user in environment *e* on link *l* (calls·seconds/hour/user) [IMT-2000]

p Index for radio environment

$P\{A\}$ Probability of the event *A* [queueing theory]

$P\{A \mid B\}$ Probability of the event A conditioned on the event B [queueing theory]

p_k Probability that the system is in state k [queueing theory]

$P_{m,n,rat,p}$ Session arrival rate per cell of service category n in service environment m for RATG *rat* in radio environment p (sessions/s/cell)

Penetration_Rate$_{e,s}$ Ratio of the number of people/vehicles subscribing to service s over the total population in environment e (%) [IMT-2000]

Population_Density$_e$ Average number of people/vehicles per unit area in environment e (users/km^2) [IMT-2000]

$Q_{m,n}$ Session arrival rate per user of service category n in service environment m (sessions/s/user)

$Q_{m,s}$ Session arrival rate per user of service s in service environment m (sessions/s/user)

R Radius of a cell (km) [IMT-2000]

rat Index for radio access technique group (RATG)

$r_{d,n,rat,p}$ Mean service bit rate of service category n in teledensity d for RATG *rat* in radio environment p (bit/s)

$r_{m,n}$ Mean service bit rate of service category n in service environment m (bit/s)

$r_{m,s}$ Mean service bit rate of service s in service environment m (bit/s)

Required_Bit_Rate_per_Cell$_{e,s,l}$ Required bit rate for service s per cell of environment e on link l (Mbit/s/cell) [IMT-2000]

Required_Channels_per_Cell$_{e,s,l}$ Number of channels for service s per cell of environment e on link l (channels/cell) [IMT-2000]

Required_Channels_per_Group$_{e,s,l}$ Number of channels for service s per group of cells in environment e on link l (channels/group) [IMT-2000]

Required_Spectrum$_{e,s,l}$ Required spectrum for service s of environment e on link l (MHz) [IMT-2000]

s Index for service [IMT-2000]

s Number of servers (dimensionless) [queueing theory]

s_n Mean packet size for service category n (bit/packet)

$s_n^{(2)}$ Second moment of the packet size for service category n (bit^2/packet^2)

Service_Channel_Bit_Rate$_{e,s}$ Service channel bit rate for service s in environment e (kbit/s) [IMT-2000]

$T_{d,n,rat,p}$ Offered traffic per cell of service category n in teledensity d for RATG *rat* in radio environment p (bit/s/cell)

$U_{m,n}$ User density of service category n in service environment m (users/km^2)

$U_{m,s}$ User density of service s in service environment m (users/km^2)

Users_per_Cell$_{e,s}$ Average number of users of service s per cell of environment e (users/cell) [IMT-2000]

W Waiting time (s) [queueing theory]

W_n Waiting time of class n (s) [queueing theory]

$w_{m,s}$ Weight for the average session duration for service s in service environment m (dimensionless)

$\bar{w}_{m,s}$ Weight for the mean service bit rate for service s in service environment m (dimensionless)

$X_{hs;m}$ Population coverage percentage for hot spot in service environment m (dimensionless)

$X_{macro;m}$ Population coverage percentage for macro cell in service environment m (dimensionless)

$X_{micro;m}$ Population coverage percentage for micro cell in service environment m (dimensionless)

$X_{pico;m}$ Population coverage percentage for pico cell in service environment m (dimensionless)

$\alpha_{e,s}$ Weighting factor for service s of environment e (dimensionless) [IMT-2000]

β Adjustment factor (dimensionless) [IMT-2000]

$\Gamma(k)$ Gamma function

$\gamma(k,z)$ Incomplete gamma function

Δ_n Maximum allowable mean delay for service category n (s/packet)

$\delta(t)$ Dirac impulse at $t = 0$

$\eta_{d,rat,p}$ Spectral efficiency in teledensity d for RATG rat in radio environment p (bit/s/Hz/cell)

λ Arrival rate in a Poisson arrival process (arrival/s) [queueing theory]

λ_n Arrival rate of class n in a Poisson arrival process (arrival/s) [queueing theory]

$\lambda_{d,n,rat,p}$ Packet arrival rate per cell of service category n in teledensity d for RATG rat in radio environment p (packets/s/cell)

μ Parameter of exponential distribution (1/s) [queueing theory]

$\mu_{m,n}$ Average session duration of service category n in service environment m (s/session)

$\mu_{m,s}$ Average session duration of service s in service environment m (s/session)

$\xi_{\text{hs};m,n}$ Intermediate traffic distribution ratio for hot spot of service category n in service environment m (dimensionless)

$\xi_{\text{macro};m,n}$ Intermediate traffic distribution ratio for macro cell of service category n in service environment m (dimensionless)

$\xi_{\text{micro};m,n}$ Intermediate traffic distribution ratio for micro cell of service category n in service environment m (dimensionless)

$\xi_{\text{pico};m,n}$ Intermediate traffic distribution ratio for pico cell of service category n in service environment m (dimensionless)

$\xi_{\text{pico\&hs};m,n}$ Intermediate traffic distribution ratio for pico cell and hot spot of service category n in service environment m (dimensionless)

$\xi_{m,n,rat,p}$ Unicast traffic distribution ratio of service category n in service environment m to RATG rat in radio environment p (dimensionless)

$\xi_{m,n,rat,p}^{\text{Multicast}}$ Multicast traffic distribution ratio of service category n in service environment m to RATG rat in radio environment p (dimensionless)

ξ_{rat} Traffic Distribution ratio among available RATGs (dimensionless)

π_n Maximum allowable blocking probability for service category n (dimensionless)

$\rho_{d,n,rat,p}$ Offered traffic per cell of service category n in teledensity d for RATG rat in radio environment p (Erlang/cell)

$\max\{\cdots\}$ Maximum operation

$\min\{\cdots\}$ Minimum operation

$:=$ Equals by definition

$\lceil x \rceil$ Ceiling function (the smallest integer greater than or equal to x)

$\lfloor x \rfloor$ Floor function (the largest integer not exceeding x)

Appendix D

ITU-R Documents and Web Sites

D.1 ITU-R Recommendations

The ITU-R Recommendations constitute a set of international technical standards developed by the Radiocommunication Sector of the ITU. They are the result of studies undertaken by Radiocommunication Study Groups, and approved by ITU Member States. M series of the ITU-R Recommendations are for mobile, radiodetermination, amateur and related satellite services. They are available at the web site 'http://www.itu.int/rec/R-REC-M/en'. Those relevant to this book are as follows.

Recommendation ITU-R M.816-1, Framework for services supported on International Mobile Telecommunications-2000 (IMT-2000), 1997.

Recommendation ITU-R M.1079-2, Performance and QoS requirements for International Mobile Telecommunications-2000 (IMT-2000) access networks, 2003.

Recommendation ITU-R M.1390, Methodology for the calculation of IMT-2000 terrestrial spectrum requirements, 1999.

Recommendation ITU-R M.1455-2, Key characteristics for the International Mobile Telecommunications-2000 (IMT-2000) radio interfaces, 2003.

Recommendation ITU-R M.1457-6, Detailed specifications of the radio interfaces of International Mobile Telecommunications-2000 (IMT-2000), 2006.

Recommendation ITU-R M.1645, Framework and overall objectives of the future development of IMT-2000 and systems beyond IMT-2000, 2003.

Recommendation ITU-R M.1651, A method for assessing the required spectrum for broadband nomadic wireless access systems including radio local area networks using the 5 GHz band, 2003.

Spectrum Requirement Planning in Wireless Communications Edited by Hideaki Takagi and Bernhard H. Walke
© 2008 John Wiley & Sons, Ltd

Recommendation ITU-R M.1768, Methodology for calculation of spectrum requirements for the future development of the terrestrial component of IMT-2000 and systems beyond IMT-2000, 2006.

D.2   ITU-R Reports

The ITU-R Report is a technical, operational or procedural statement, prepared by a Study Group on a given subject related to a current Question. M series of the ITU-R Reports are for mobile, radiodetermination, amateur and related satellite services. They are available at the web site 'http://www.itu.int/publ/R-REP-M/en'. Those relevant to this book are as follows.

Report ITU-R M.2023, Spectrum requirements for International Mobile Telecommunications-2000, 2000.

Report ITU-R M.2072, World mobile telecommunication market forecast, 2006.

Report ITU-R M.2074, Radio aspects for the terrestrial component of IMT-2000 and systems beyond IMT-2000, 2006.

Report ITU-R M.2078, Estimated spectrum bandwidth requirements for the future development of IMT-2000 and IMT-Advanced, 2006.

Report ITU-R M.2079, Technical and operational information for identifying spectrum for the terrestrial component of future development of IMT-2000 and IMT-Advanced, 2006.

Report ITU-R M.2109, Sharing studies between IMT Advanced systems and geostationary satellite networks in the fixed-satellite service in the 3400–4200 and 4500–4800 MHz frequency bands, 2007.

Report ITU-R M.2110, Sharing studies between radiocommunication services and IMT systems operating in the 450–470 MHz band, 2007.

Report ITU-R M.2112, Compatibility/sharing of airport surveillance radars and meteorological radar with IMT systems within the 2700–2900 MHz band, 2007.

D.3   Other ITU-R Documents

ITU-R Radio Regulations.
    http://www.itu.int/publ/R-REG-RR/en

'SPECULATOR', Tool for estimating the spectrum requirements for the future development of IMT-2000 IMT-Advanced.
    http://www.itu.int/ITU-R/study-groups/docs/speculator.doc

Administrative Circular Letter CACE/326, Questionnaire on the services and market for the future development of IMT-2000 and IMT-Advanced.
    http://www.itu.int/md/R00-CACE-CIR-0326/en

Resolution ITU-R 56, Naming for International Mobile Telecommunications, 2007.
http://www.itu.int/publ/R-RES-R.56-2007/en

ITU-R CPM07-2, Report of the CPM to WRC-07, 2007.
http://www.itu.int/md/R07-CPM-R-0001/en

D.4 Web Sites

- Overview of ITU
  http://www.itu.int/aboutitu/overview

- M series of ITU-R Recommendations
  http://www.itu.int/rec/R-REC-M/en

- M series of ITU-R Reports
  http://www.itu.int/publ/R-REP-M/en

- Frequency allocation chart in the United States
  http://www.ntia.doc.gov/osmhome/allochrt.pdf

- Frequency allocation chart in Finland
  http://www.ficora.fi/englanti/document/Use_of_radio_spectrum.pdf

- Frequency allocation chart in Japan
  http://www.tele.soumu.go.jp/e/search/myuse/use0303/index.htm

Bibliography

Abramowitz M and Stegun IA 1972 *Handbook of Mathematical Functions with Formulas, Graphs, and Mathematical Tables*. New York, Dover Publications.

Azuma M 2007 *Study on Management and Control of Communication Networks: Failure Management/Security Management/Transport Control*. PhD thesis, University of Tsukuba, Japan.

Brockmeyer E, Halstrom HL and Jensen A 1948 The life and works of A. K. Erlang. *Transactions of the Danish Academy of Technical Sciences*, vol. 2.

Cobham A 1954 Priority assignments in waiting line problems. *Operations Research* **2**(1), 70–76.

Cooper RB 1991 *Introduction to Queueing Theory*, 2nd edn. Amsterdam, North-Holland.

COST 231, Urban transmission loss models for mobile radio in the 900- and 1800 MHz bands. COST 231 TD(90)119 Rev. 2. The Hague, The Netherlands, September 1991.

Fortet R and Grandjean Ch 1964 Congestion in a loss system when some calls want several devices simultaneously. *Electrical Communications* **39**(4), 513–526.

Fujiki M and Gambe E 1980 *Teletraffic Theory*, Maruzen, Tokyo (in Japanese).

Gross D and Harris CM 1998 *Fundamentals of Queueing Theory*, 3rd edn. John Wiley & Sons.

Hata M 1980 Empirical formula for propagation loss in land mobile radio services. *IEEE Transactions Vehicular Technology* **VT-29**(3), 317–325.

Irnich T *A New Methodology for Radio Spectrum Requirement Prediction of Wireless Communication Systems*. PhD thesis, RWTH Aachen University, Germany, expected in 2008.

Irnich T and Walke B 2004 Spectrum estimation methodology for next generation wireless systems. In *15th IEEE Annual International Symposium on Personal Indoor and Mobile Radio Communication* (PIMRC 2004) Barcelona, Spain, 5–8 September 2004, pp. 1957–1962.

Irnich T and Walke B 2005 Spectrum estimation methodology for next generation wireless systems: Introduction and results of application to IMT-2000. In *16th IEEE Annual International Symposium on Personal Indoor and Mobile Radio Communication* (PIMRC 2005) Berlin, Germany, 11–14 September 2005, pp. 2801–2809.

Irnich T, Walke B and Takagi H 2005 System capacity calculation for packet-switched traffic in the next generation wireless systems, Part I: nonpreemptive priority queueing model for IP packet transmission. In *19th International Teletraffic Congress* (ITC-19) Beijing, China, 29 August–2 September 2005, pp. 13–23.

IST-MIND D3.3. Methodologies to identify spectrum requirements for systems beyond 3G, 2002.

IST-2003-50781 WINNER D6.2. Methodology for estimating the spectrum requirements for 'further developments of IMT-2000 and systems beyond IMT-2000', 2005.

Iversen B *Teletraffic Engineering Handbook*, ITU-D Study Group 2 Question 16/2, http://www.com.dtu.dk/tetetraffic/handbook/telenookpdf.pdf

Kaufman JS 1981 Blocking in a shared resource environment. *IEEE Transactions on Communications* **COM-29**(10), 1474–1481.

Kesten H and Runnenburg J Th 1957 Priority in waiting line problems. I and II. *Proceedings of the Koninklijke Nederlandse Akademie van Wetenschappen*, Series A, **60**, pp. 312–324 and pp. 325–336.

Kleinrock L 1975 *Queueing Systems*, vol. 1: *Theory*. John Wiley & Sons.

Kleinrock L 1976 *Queueing Systems*, vol. 2: *Computer Applications*. John Wiley & Sons.

Lott M and Scheibenbogen M 1997 Calculation of minimum frequency separation for mobile communication systems. In *Proceedings of the 2nd European Personal Mobile Communications Conference* (EPMCC '97), Bonn, Germany, pp. 133–137.

Matinmikko M 2007 *Estimation of Spectrum Requirements of IMT-Advanced Systems*. Licentiate thesis, Department of Electrical and Information Engineering, University of Oulu, Finland, 118 pages.

Mohr W 2003 Spectrum demand for systems beyond IMT-2000 based on data rate estimates. *Wireless Communications and Mobile Computing* 3(7), 817–835.

Motorola Radio Research Laboratory. An etiquette for sharing multimedia radio channels. European Telecommunications Standards Institute, ETSI RES 10/TTG-94/98, September 1994.

Okumura Y, Ohmori E, Kawano T and Fukuda K 1968 Field strength and its variability in VHF and UHF land-mobile radio service. *Review of the Electrical Communication Laboratory* 16 (9–10), 825–873.

Roberts JW 1981 A service system with heterogeneous user requirements – Application to multi-services telecommunication systems. In *Performance of Data Communication Systems and their Applications* (ed. Pujolle G). North-Holland, 1981, pp. 423–431.

Shortle JF and Brill PH 2005 Analytical distribution of waiting time in the M/iD/1 queue. *Queueing Systems* 50(2), Nos. 2/3, 185–197.

Stevens WR 1994 *TCP/IP Illustrated*, vol. 1: *The Protocols*. Addison-Wesley.

Syski R 1986 *Introduction to Congestion Theory in Telephone Systems*, 2nd edn. Elsevier.

Takagi H 1991 *Queueing Analysis A Foundation of Performance Evaluation*, vol. 1: *Vacation and Priority Systems*. Elsevier.

Takagi H, Yoshio H, Matoba N and Azuma M 2006 Methodology for calculation of spectrum requirements for the next generation mobile communication systems. *The IEICE Transactions on Communications* **J89-B**(2), 135–142 (in Japanese).

Thompson K, Miller G and Wilder R 1997 Wide area Internet traffic patterns and characteristics. *IEEE Network Magazine* 11(6), 10–23.

Walfisch J and Bertoni H 1998 A theoretical model of UHF propagation in urban environments. *IEEE Transactions on Antennas and Propagation* **AP-36**(12), 1788–1796.

Walke B 2002 *Mobile Radio Networks – Networking, Protocols and Traffic Performance*, 2nd edn. John Wiley & Sons.

Walke B and Kumar V 2003 Spectrum issues and new air interfaces. *Computer Communications* 26(1), 53–63.

Walke B, Mangold S and Berlemann L 2006 *IEEE 802 Wireless Systems: Protocols, Multi-Hop Mesh/Relaying, Performance and Spectrum Coexistence*. John Wiley & Sons.

Withers D 1999 *Radio Spectrum Management*, 2nd edn. IEEE Press.

Wolff RW 1989 *Stochastic Modeling and the Theory of Queues*. Prentice-Hall.

3rd Generation Partnership Project (3GPP) Technical specification group services and system aspects; Quality of service (QoS) concept and architecture, Release 6, 3GPP TS 23.107 V6.4.0, March 2006.

Index

Activity factor, 54
Adjustment factor, 68
Application data rate, 84, 86, 106, 121, 137
Application service, 47
Arrival rate, 90, 94, 202, 204, 207, 211, 215
Average session duration, 80, 85, 121, 127, 152, 219

Background, 79, 150
Bearer service, 47
Best efforts, 79
Birth-and-death process, 202, 204, 207
Blocking probability, 60, 93, 203, 210, 212
Busy hour call attempts (BHCA), 54

Calculation tool, 76, 101, 219, 242
Call duration, 54
Cardano's formula, 96
Cell area, 51, 83, 137
Central business district (CBD), 47
Channel capacity, 95
Circuit-switched service, 48, 60, 91, 92
Cobham's formula, 95, 218
Conversational, 79, 150

Degenerated hyperexponential distribution, 178
Delay, 168, 180
 distribution, 181
 mean, 95
 percentile, 167
Dense urban, 82
Density, 46
Dirac impulse, 170, 178, 181
Distribution function (DF)
 delay, 168, 181, 183
 waiting time, 172, 178, 179, 182
Downlink, 48, 97, 219

Environment, 46
Erlang, 59, 91
Erlang-B formula, 60, 203
Erlang-C formula, 62, 206
Example calculation, 88, 129, 142, 149
Exponential distribution, 178, 201, 202, 204, 206, 207, 211

First-come first-served (FCFS), 95, 206, 215

Flexible spectrum use, 17, 84, 98
Flow chart
 delay percentile, 169
 market data analysis, 123
 methodology for IMT-2000, 50
 methodology for IMT-Advanced, 78

Gamma distribution, 179
Global balance equations, 208
Group size, 55
Guard band between operators, 68, 84, 98, 141

High interactive multimedia, 47
High multimedia, 47, 79
High-density, 46
Hot spot, 83
Hyperexponential distribution, 178

In-building, 46
Insensitivity, 204
Interactive, 79, 150

Kaufman–Roberts algorithm, 93, 214
Kendall's notation, 201
Kleinrock's conservation law, 218

Laplace–Stieltjes transform
 delay, 168
 waiting time, 171
Little's law, 207, 217
Local balance equations, 209
Low multimedia, 79
Low rate data, 79

$M/\{iD\}/1$, 172
$M/D^*/1$, 170
$M/G/1$ nonpreemptive priority queue, 95, 171, 215
$M/G/1$-FCFS, 171
$M/M/s$ delay system, 61, 204
$M/M/s/s$ loss system, 60, 202, 207
Macro cell, 83
Market data, 113, 151, 219
Market setting, 105, 130, 151–154
Maximum allowable
 blocking probability, 82, 92, 150
 mean packet delay, 82, 94, 150